Hierarchies in the Brain, Mind and Behaviour

Gerald Wiest

Hierarchies in the Brain, Mind and Behaviour

A Principle Of Neural and Mental Function

Foreword by Mark Solms

palgrave
macmillan

Gerald Wiest
Department of Neurology
Medical University of Vienna
Vienna, Austria

ISBN 978-3-031-78029-5 ISBN 978-3-031-78030-1 (eBook)
https://doi.org/10.1007/978-3-031-78030-1

The original submitted manuscript has been translated into English. The translation was done using artificial intelligence. A subsequent revision was performed by the author(s) to further refine the work and to ensure that the translation is appropriate concerning content and scientific correctness. It may, however, read stylistically different from a conventional translation.
Translation from the German language edition: "Hierarchien in Gehirn, Geist und Verhalten Ein Prinzip neuraler und mentaler Funktion" by Gerald Wiest, © Springer-Verlag Vienna 2009. Published by Springer Nature. All Rights Reserved.

This Palgrave Macmillan imprint is published by the registered company Springer Nature Switzerland AG.
The registered company address is: Gewerbestrasse 11, 6330 Cham, Switzerland

The heart has its reasons that reason does not know.
Blaise Pascal, Pensées IV, 277

To Sylvia and Nicholas

Foreword

It is a peculiarly pleasant task to introduce a book which deals with a subject so close to my heart that I would like to have written it myself.

With this book, Gerald Wiest brings to contemporary awareness a largely unrecognised but very important fact about the history of behavioural neuroscience, namely that it has from the start given rise to two parallel traditions. These might be called the 'modular' and 'dynamic' traditions respectively. The first of these traditions—which has always been the dominant one—had its origins in classical localisationism (and before that, in phrenology), the theoretical assumptions of which persist today in the way that cognitive neuroscientists interpret functional magnetic resonance images. The other 'dynamic' tradition has its origins in the hierarchical conceptions of Jackson and continues to this day in the deeply evolutionary perspective that informs the affective neuroscience of Panksepp, and also the nascent interdiscipline now known as neuropsychoanalysis.

Anyone who respects reality must acknowledge that the life of the mind is a complex and difficult subject, which does not sit easily with the reductive simplicity of modularism, however appealing that simplicity may be from the purely didactic point of view. (This was the basis of Freud's early dismissal of classical behavioural neurology as 'a silly game of permutations'.) The alternative, dynamic view is far more difficult to grasp, and to teach, and this must be one of the main sources of the ongoing resistance to it—if not downright ignorance of it. This way of thinking about the mind is much more demanding intellectually than the notion that complete mental functions can be assigned to individual structures or 'modules'.

Ultimately, however, the difficulty and subtlety of the dynamic conception is also what makes it convincing. Who can really sustain a view of mental life that denies it the dimension of depth? This is true not only in the face of the actual clinical and developmental phenomena, of neurology and psychiatry alike, but also—in fact especially—in light of the ocean of evidence upon which all of evolutionary biology rests. How can the evolutionary Weltanschauung ever be reconciled with a modular theory? Mental functions simply must carry within themselves their maturational and evolutionary trajectory, which necessarily implies that structure/function relationships in the existing phenotype cannot convey the full story. The fuller story is inevitably submerged in earlier and deeper relationships, in layers of both ontogenetic and phylogenetic organisation that are not so superficially apparent.

Gerald Wiest's tactful and modest treatment of this profound subject matter encourages contemporary mental scientists to strive once more to transcend the arrogance of the present—not only with respect to the biological prehistory of the adult human mental apparatus but also with respect to the rich theoretical heritage of behavioural neuroscience itself.

Mark Solms

Department of Neuropsychology,
Neuroscience Institute,
University of Cape Town,
Cape Town, South Africa

Preface

The idea of a hierarchical organisation of the human mind can be traced back to Plato's conceptions of the structure of the soul. This book aims to highlight hierarchical principles of the brain and mind, and trace their presence in various contemporary scientific disciplines. The arc spans from the philosophical approaches of Herbert Spencer and his ideas about the hierarchical organisation of the brain, to Hughlings Jackson, who made this the basis for his doctrine of neurology. Jackson, in turn, influenced Sigmund Freud in his conception of the functioning of the mental apparatus. With the introduction of the Triune Brain Theory, Paul MacLean created a neurobiological model of the brain that views behaviour from hierarchical and phylogenetic perspectives, and represents a milestone among the concepts of evolutionary neuroethology. Finally, the theory of microgenesis is discussed, which represents a theory of neural and mental function based on hierarchical principles. The perspective of the brain and mind from hierarchical viewpoints allows for explanatory models for phenomena and aspects of human experience and behaviour that are not guaranteed by other views.

I would like to thank the following people who have contributed to the creation of this book in various ways: Eduard Auff, Robert W. Baloh, Peter Basham, Christina Bauer, Christoph Baumgartner, Thomas Berger, Andrea Binder, Elisabeth Brainin, Olaf Breidbach, Emma Butterfield, Lüder Deecke, Christine Diercks, Irenäus Eibl-Eibesfeldt, Renate Fanta, Alberto Ferrús, Georg Fodor, Tatjana Gawron, Susanne Habermann, Michael G. Halmagyi, Silvia Hildebrand, Ruth Koblizek, Vera Ligeti, Christian Müller, Jaak Panksepp, Adam J. Perkins, Bettina Reiter, Kees

Rookmaaker, Helen Sanders, Barbara Schmidt-Loeffler, Karola Schrader, Dagmar Schreiber, Tatjana Sepin-Strünkelnberg, Robert Silvers, Mark Solms, Samy Teicher, Oliver Turnbull, Véronique Van de Ponseele, John van Wyhe, Nicole Welz, Rudolf Wenger, Christian Wiest. My special thanks go to Ms. Mag. Eichhorn and Ms. Akbaba from Springer Verlag, who were always there to help and advise me.

In addition, I would like to thank the following institutions for their support and assistance: Cambridge University; Darwin Online, Cambridge; Ernst Haeckel House, Jena; Research Centre for Human Ethology, Max Planck Society; Instituto Cajal, Madrid; Institute for the History of Medicine—Josephinum, Vienna; Medical University of Vienna; National Museum of Natural History, Paris; National Portrait Gallery, London; Royal College of Physicians of London; Sigmund Freud Society, Vienna; Sigmund Freud Private Foundation, Vienna; University of California Library, Los Angeles (UCLA), USA; Runaway Technology INC, Marshfield, USA.

Vienna, in the spring of 2009

Gerald Wiest

Acknowledgement

Printed with the support of the *Federal Ministry of Science and Research in Vienna*.

Contents

List of Figures

List of Tables

CHAPTER 1

The Biological and Philosophical Foundations of a Theory

According to the views of modern biology, the organic world differs from the inorganic primarily through the development of hierarchies (Mayr 1997). Hierarchical biological principles are therefore found both in ontogeny and in phylogeny. Current ideas about the evolution of the brain correlate with current concepts of general evolution. Accordingly, (brain) evolution starts from an early, simple state, which increasingly complicates into several blueprints. Further development is characterised by branching or radiation and constancy, as well as by increasing complexity (occasionally also by simplifications) (Roth and Wullimann 1996).

The evolutionary development of the brain and mind is ultimately due to the spatial and temporal variability of vital resources and life-threatening dangers. Complex brains are found in living beings that have a continuous high energy requirement, or in organisms that only become capable of reproduction late in their development. These correlations show that the advent of the theory of evolution marked a decisive turning point in understanding the structure and function of the brain.

1.1 The Beginnings of the Theory of Evolution

The pioneer of the theory of evolution, Jean Baptiste de Lamarck (1744–1829), saw in the nervous system a crucial key to understanding evolutionary processes. Lamarck, who initially studied medicine and then

G. Wiest, *Hierarchies in the Brain, Mind and Behaviour*,
https://doi.org/10.1007/978-3-031-78030-1_1

increasingly turned to botany, chemistry, meteorology and finally zoology, never used the term "evolution" himself, but rather spoke of the "path that nature took in the creation of all living organisms".

Lamarck, who had been working at the Paris Muséum d`Histoire Naturelle as a "Professor of Insects and Worms" since 1794, first referred to his theory of evolution in 1800 in the context of an opening lecture for his course on invertebrates, the draft of which was published the following year (Lamarck 1801). Lamarck postulated at this point that changed environmental conditions and physical needs in animals might possibly lead to new habits, which would result in a "strengthening" of certain body parts or organs through frequent repetition. As a result, new body parts or organs would gradually emerge, as the acquired modifications would be passed on to the following generations through reproduction. In 1802, a decisive expansion in Lamarck's concept of evolution was made with the publication of his work *Recherches sur l'organisation des corps vivans* (Lamarck 1802), in which he dealt with the increasing degrees of complexity in the animal kingdom. The latter Lamarck explains by assuming a natural tendency of the organic to develop from the simplest to the most complex. This universal principle of the organic formed an ideal complement to his theory of the mineral kingdom, which was subject to the opposite natural tendency (Lamarck was of the opinion that all substances of the mineral kingdom ultimately arose through the disintegration of organic remains). Lamarck carried out the most detailed presentation of his evolutionary theories in the work *Philosophie zoologique*. After a more detailed argument regarding the increasing degrees of complexity in the plant and animal kingdom, Lamarck seeks a physical explanation for the emergence of higher cognitive functions in the third part of the treatise (Lamarck 1809). Materialists of the eighteenth century had previously circumvented the question of the origin of the mind insofar as they assumed that thinking was inherent in matter. Lamarck's revolutionary idea in this context was the theory that the progressive development of higher cognitive functions correlates with the structural development of the nervous system. Due to his botanical and zoological studies, Lamarck already had comprehensive explanatory models for the evolution of new structures and systems, which he now also applied to the nervous system. In his view, higher mental functions could only arise as a result of the increasing structural complexity of the brain.

In the concept of "inner sensation" (sentiment intérieur), Lamarck saw a determining factor for the functioning and evolution of higher developed animals. He differentiated between the voluntary motor skills of higher developed animals, which could lead to certain habits (such as in the search for food or the avoidance of danger) and involuntary movement, which he called "reflex movements". According to this concept, an animal's brain with a "inner sensation" (such as hunger) would induce voluntary movements to satisfy the need. If these actions were now often repeated, new organs or new body parts could result.

A sudden, strong stimulus, such as a loud noise, would induce a reflex movement. Lamarck understood his concept of "sentiment intérieur" as an important key to understanding many functions of the nervous system, such as instinct, will, memory, understanding and judgement.

The best-known summary of Lamarck's evolutionary theories can be found in his work *Histoire naturelle des animaux sans vertèbres* (Lamarck 1815–1822), in which the concept of evolution is presented in four laws. The first law refers to the principle of increasing organic complexity in the animal and plant kingdom. The second law addresses the development of new organs through the indirect influence of the environment on an animal. The third law, the principle of use and disuse, explains the change of body parts of an organism through the occurrence of new habits. The assumption of a slow and gradual evolution implied the fourth law of Lamarck, namely the postulate of the inheritance of acquired characteristics, without which the emergence of new structures could not have been explained.

Lamarck was more of a natural philosopher than a scientist throughout his life. From the beginning of his studies, he was more interested in general principles in nature than in details. Consequently, he often neglected to substantiate his evolutionary concepts with concrete examples. Although his most important postulate of the inheritance of acquired characteristics is now considered obsolete, Lamarck still deserves an important place in the history of science, not only because of his pioneering work in botany or zoology but above all because of the creation of the first coherent theory of evolution. However, it was only the theory of evolution by Charles Darwin that secured Lamarck's place in the history of science.

1.2 Charles Darwin's Theory of Evolution

The decisive breakthrough to the modern, still valid theory of evolution was only achieved by Charles Darwin (1809–1882). Darwin's ideas not only revolutionised all of biology but also fundamentally changed man's self-understanding (Fig. 1.1).

As a medical student, Darwin was introduced to Lamarck's theories of evolution by the zoologist Robert Edmond Grant at the University of Edinburgh in 1827. After starting his theology studies in Cambridge, Darwin became increasingly interested in botany, zoology and palaeontology. The years 1831–1836, which Darwin finally spent as a naturalist on the survey ship "Beagle" of the British Admiralty, were not only the most important event of his scientific life but also represent a turning point in the history of biology. When the "Beagle" left England in 1831, Darwin was still of the belief that each species had been individually created. However, due to his extensive botanical and zoological observations, Darwin increasingly doubted the immutability of species, which he only noted in his notebook "Transmutation of Species" in 1837 (Darwin 1837) after the end of the journey. Based on his observations, he concluded that

Fig. 1.1 Charles Darwin (1809–1882): Reproduced with permission from John van Wyhe ed. (2002-). (The Complete Work of Charles Darwin Online. (http://darwin-online.org.uk/))

probably all living beings on earth, including humans, could be traced back to common ancestors.

For Darwin, this assumption was the answer to many of his previously unexplained observations, such as the fact that embryos of lizards, chickens or rabbits resemble each other, but the adult animals differ significantly from each other or that many animals only have rudimentary organs. But also behavioural similarities between animals and humans, such as yawning in horses and humans, or the similarities in emotional reactions of humans and orangutans suddenly made sense to Darwin. At this point, he already saw evolution as a process of adaptation of species to the environment, without recognising the guiding mechanism behind it. The solution to this problem came to him by chance in 1838, after reading the book *Essay on the Principle of Population* by Thomas Robert Malthus (Malthus 1798). Darwin applied the ideas about "population laws" described in Malthus's book to the animal and plant kingdom, and thus discovered the principle of "natural selection", which was to make him the founder of modern evolutionary theory. Only through the urging of friends did he begin to write down his theory of evolution in 1856. In 1858, Darwin received from Alfred Russel Wallace—a naturalist working in Southeast Asia—a letter that presented a nearly identical treatise to Darwin's theories but could not rely on such a rich body of evidence as Darwin's. Darwin and Wallace finally presented their works together in July 1858 before the Linnean Society in London. The publication of Darwin's main work *On the Origin of Species*, which Darwin modestly referred to as an "abstract", appeared on 24 November 1859 in an edition of 1250 copies and was sold out on the same day (Darwin 1859). In this book the five basic principles of his evolutionary theory, which he worked on for more than 20 years, are summarised as follows:

- Evolution is based on common descent, with development not occurring linearly but in the form of branching stems, as can be seen in the diversity of organisms.
- Organisms tend to constantly evolve.
- Species multiply over time.
- Evolution occurs in the form of a slowly progressive change.
- The mechanism of evolution is based on competition among numerous individuals for limited resources, with a fundamental overproduction of offspring among organisms.

In his work *The Descent of Man*, published in 1871, Darwin first focused on the similarities of physical characteristics in humans and primates, then pointed out the common psychological and physiological processes involved in mating, reproduction, menstruation, birth and childhood (Darwin 1871). Darwin was, alongside Lamarck, the first scientist to attribute the development of the brain and mind to an evolutionary process. He argued that animals and humans share the urge for self-preservation, maternal care, but also pleasure, arousal, imitation or memory, and that even the emergence of moral feelings can be explained by evolution. In his opinion, morality would arise in any creature equipped with social instincts through an evolutionary process, as soon as the intellectual abilities for it were developed. Subsequently, Darwin increasingly turned to questions of the evolution of the brain and behaviour. In his book *The Expression of the Emotions in Man and Animals*, he dealt with the connections between facial expression and emotional behaviours in humans and animals (Darwin 1872). Darwin thus made an important contribution to psychology and is considered a pioneer of ethology (Fig. 1.2).

One of the first advocates of Darwin's theory of evolution in the German-speaking world was Ernst Haeckel (1834–1919), a professor of zoology at the University of Jena. Haeckel coined the terms ontogenesis, the development from the fertilised egg to the mature organism, and phylogenesis, the "history of the development of the tribes" (Haeckel 1866). Haeckel repeatedly emphasised the parallels between onto- and phylogenesis and formulated the "Biogenetic Law" (Fig. 1.3), according to which an organism recapitulates the entire phylogenesis in its ontogenesis (Haeckel 1874).

Although several aspects of Haeckel's explanations of the law are no longer tenable today, the basic principle he discovered is still valid and is confirmed by current molecular biological research.

1.3 Evolution and Dissolution

The British philosopher Herbert Spencer (1820–1903) was the first to introduce the concept of evolution in the modern sense and expanded evolutionary principles in later works to his concepts of society and the universe. Already in his youth, then still an employee of a railway company, he developed a pronounced interest in evolution, after repeatedly coming across fossils during excavation work.

Fig. 1.2 Illustration from Charles Darwin's book *The Expression of the Emotions in Man and Animals* (1872). (Reproduced with permission from John van Wyhe ed. (2002-). The Complete Work of Charles Darwin Online. (http://darwin-online.org.uk/))

A few years later, Spencer became an enthusiastic advocate of Lamarckism, to finally present his concept of an evolutionary psychobiology in the *Principles of Psychology* (Spencer 1855). In this work, Spencer deals extensively with the structure, function and evolution of the nervous system. According to his conception, the mind can only be understood by understanding its phylogenetic development, with the evolution of consciousness going hand in hand with the evolution of the nervous system.

Fig. 1.3 Illustration of the "Biogenetic Law" by Ernst Haeckel from *Anthropogenie oder Entwicklungsgeschichte des Menschen* (1874). (The illustration shows the morphological similarities of embryos of different species (from left to right: fish, salamander, turtle, bird, pig, cattle, rabbit and human))

Although a phylogenetic perspective proves for Spencer that animals constantly adapt their "internal conditions" to the external conditions of the environment, Spencer also adheres to the Lamarckian view of the heritability of acquired characteristics. According to Spencer, the phylogenesis of consciousness proves a universal law of evolution: The development from an undifferentiated, incoherent homogeneity to a differentiated, coherent heterogeneity. According to this conception, nerve centres are initially hardly organised, and the stimulus or information processing is slow and laborious. With frequent use, these nerve centres would subsequently become more organised, complex, and—due to the increasing connections between individual elements—more differentiated and coherent. To illustrate this evolutionary principle, Spencer refers to an example of ontogenesis: Language is learned by the toddler only slowly and laboriously, which, according to Spencer, points to the undifferentiated nature of the responsible nerve centres. However, as age increases, speaking becomes easier, more articulated and almost automatic, which is attributed to the increasing differentiation and high organisation of the matured nerve centres. Although Spencer only theoretically dealt with the nervous system, he nevertheless had concrete ideas about the structure and function of the brain. He conceived the evolutionary change of the nervous system to higher complexity as the superimposition of ever newer nerve associations, with pre-existing associations increasingly coordinating their activities. The anatomical correlate of these superimpositions are new layers of nerve cells (Fig. 1.4).

For Spencer, therefore, the structures of the central nervous system not only reflect the experiences of the individual's past but also those of the

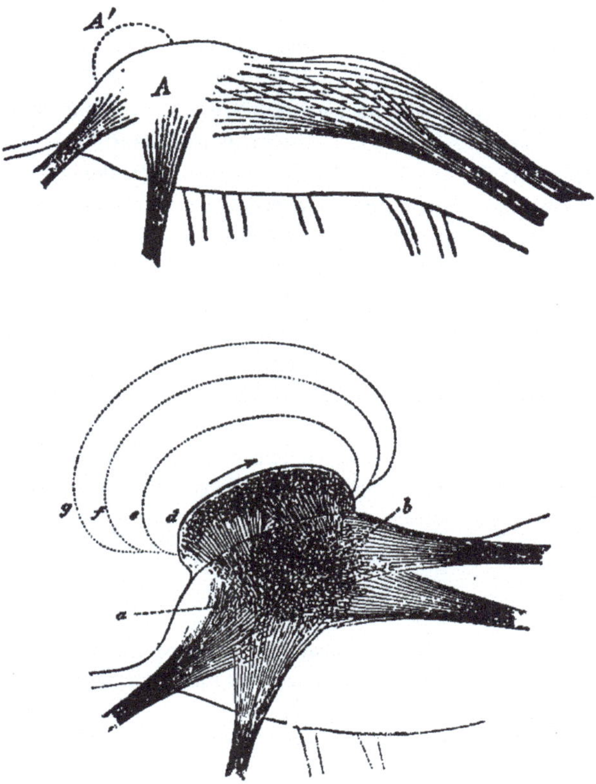

Fig. 1.4 Herbert Spencer's ideas about the evolutionary change of the nervous system through superimposition of nerve associations in invertebrate ganglia. (From the original ganglion (A) a new structure (A′) is formed by superimposition. The excitation process no longer proceeds from a to b, but via the new neural layers (d, e, f, g). From *The Principles of Psychology*, 3rd Ed. (Spencer 1896). Copyright: Springer)

ancestors. Even at the end of his life, Spencer remained convinced that the inheritance of acquired characteristics is an implicit component of evolution. At the end of psychological evolution, according to Spencer's concept, is the ability to form the most complex neural representations in response to external stimuli. The principles of psychobiological evolution are effective at all levels, so that even abstract ideas and highly developed moral concepts differ only in the extent of automatic contractions of microorganisms. In a phylogenetic series, according to Spencer, older origins of consciousness are increasingly transformed into reflex centres, so that their archaic forms of consciousness are no longer accessible to the more developed centres. Spencer brings this important aspect of his evolutionary theory to the point in the *Principles of Psychology*: "Beyond the limits of the coherent aggregate of activities ... constituting consciousness, there exist other activities of the same intrinsic nature, which being cut off are rendered foreign to it" (Spencer 1855).

The application of the fundamental evolutionary principle propagated by Spencer to the nervous system—namely the development from an incoherent homogeneity to a complex, interconnected heterogeneity—implies that a lesion at a higher cerebral level reveals behaviours and forms of consciousness that originate from an earlier evolutionary stage. This thought would later form a basis for the development of neurology as a medical science and have a decisive influence on the intellectual history of the twentieth century.

Bibliography

Darwin CR (1837) Notebook B (Transmutation of species).

Darwin CR (1859) On the origin of species by means of natural selection. John Murray, London.

Darwin CR (1871) The descent of man, and selection in relation to sex. John Murray, London.

Darwin CR (1872). The expression of the emotions in man and animals. John Murray, London.

Haeckel E (1866) Die Generelle Morphologie der Organismen. Berlin, G. Reimer.

Haeckel E (1874) Anthropogenie oder Entwicklungsgeschichte des Menschen. Engelmann, Leipzig.

Lamarck JB (1801) Systéme des animaux sans vertébres précédé du „Discours d'ouverture du cours de zoologie de l'an VIII" (Paris, an IX).

Lamarck JB (1802) Recherches sur l'organisation des corps vivans précédé du „Discours d'ouverture du cours de zoologie de l'an X" (Paris, an X).

Lamarck JB (1809) Philosophie zoologique. 2 vols, Paris.

Lamarck JB (1815–1822) Histoire naturelle des animaux sans vertèbres. 7 vols, Paris.

Malthus TR (1798) An essay on the principle of population. Johnson J, St. Paul´s Church Yard, London.

Mayr E (1997) This is biology. Harvard University Press, Cambridge.

Roth G, Wullimann MF (1996) Evolution der Nervensysteme und der Sinnesorgane. In: Dudel J, Menzel R, Schmidt RF (Eds) Neurowissenschaft. Vom Molekül zur Kognition. Springer Verlag, Berlin, Heidelberg, New York, pp 1–31.

Spencer H (1855) The principles of psychology. London: Longman, Brown, Green, and Longmans.

Spencer H (1896) The principles of psychology, 3rd edition, 2 volumes in 3. Appleton, New York.

CHAPTER 2

Hierarchical Principles in the Nervous System

2.1 Evolution and Dissolution of the Nervous System

John Hughlings Jackson, who would later go down in the history of medicine as one of the founding fathers of clinical neurology, was significantly influenced by the ideas of Herbert Spencer. At the beginning of his career as an assistant at the London *National Hospital for the Paralysed and the Insane*, he was impressed by the sometimes strange symptoms of neurological patients.

In search of possible explanations for these phenomena, Jackson found in Spencer's writings, especially in his work *Principles of Psychology* (Spencer 1855), the key to understanding. The principle that Jackson applies to his neurological observations and thus extends to the function of the entire nervous system is that of evolution and dissolution, where he particularly develops a science of diseases of the nervous system from the concept of dissolution—the reversal of evolution (Taylor 1931/32).

According to Spencer's theory, neurological symptoms such as hemiplegia, aphasia or epilepsy represent a dissolution of the nervous system caused by brain lesions. From this perspective, cerebral failures (dissolution) provide an insight into the phylogenetic development (evolution) of the brain. For Jackson, evolution represents a process from a state of highest to lowest organisation, where he uses the term degree of organisation

G. Wiest, *Hierarchies in the Brain, Mind and Behaviour*, https://doi.org/10.1007/978-3-031-78030-1_2

to refer to the extent of the preformed automatic connections of the neural units with each other. The spinal reflex arc as an example of a phylogenetically simple but highly organised and less complex unit thus contrasts with the most developed cortical centres, which are the least organised (in Jackson's words the "most helpless centres") but structurally represent the most complex units (Fig. 2.1).

At the same time, however, evolution also represents a turn from a state of least to highest modifiability. If, for example, the highest cerebral centres were not modifiable, human behaviour would correspond to that of a simple machine, since no new receptions could be made. A modifiability

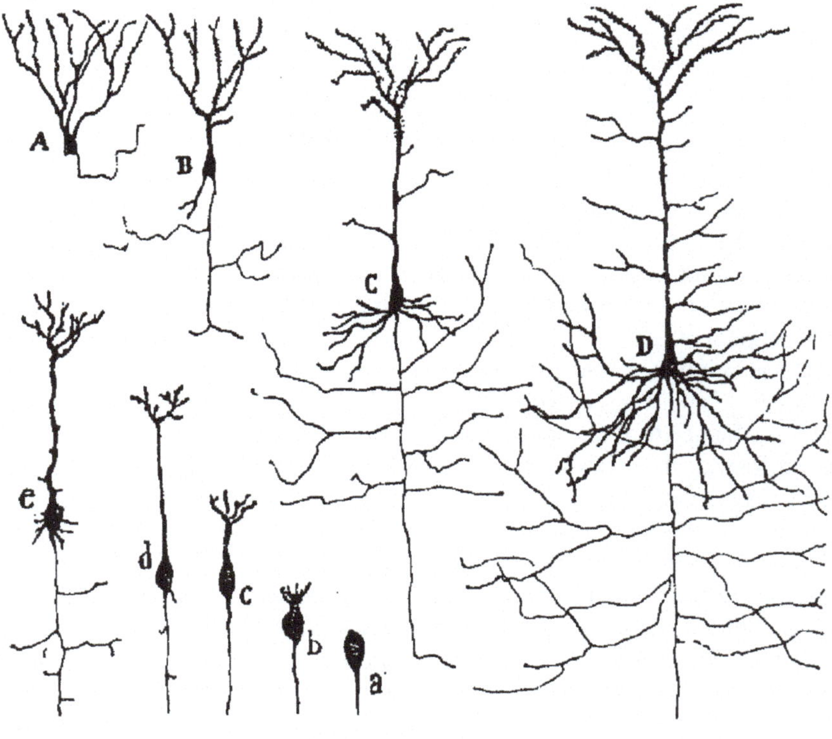

Fig. 2.1 Representation of the phylogenetic (ABCD) and ontogenetic (abcde) development of pyramidal cells (A-Frog, B-Lizard, C-Rat, D-Human) by Ramón y Cajal (Ramón y Cajal 1892, 1909–1911). (With kind permission from Prof. Alberto Ferrús, Instituto Cajal, Madrid)

of the simplest or deepest (vital) centres—similar to those of the highest—would, on the other hand, be incompatible with life. In this respect, Jackson also agrees with Spencer's ideas that the deeper brain structures are responsible for automatisms or reflex behaviour, while the highest centres serve higher functions such as learning and memory. In functional terms, Jackson saw the nervous system, from its simplest structure (the reflex arc) to the highest cortical areas, always obeying a sensorimotor mechanism (Fig. 2.2).

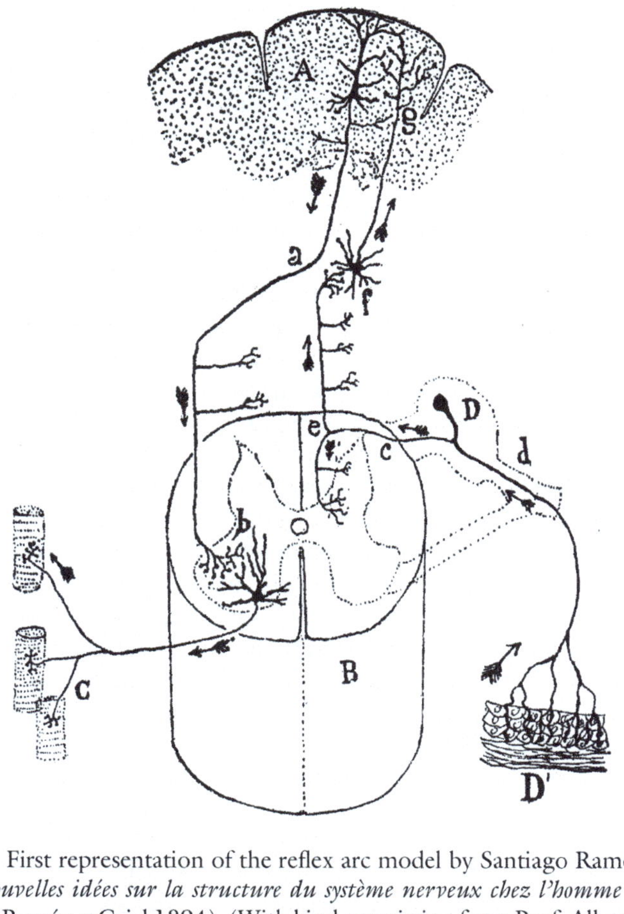

Fig. 2.2 First representation of the reflex arc model by Santiago Ramón y Cajal in "*Les nouvelles idées sur la structure du système nerveux chez l'homme et chez les vertebré*" (Ramón y Cajal 1894). (With kind permission from Prof. Alberto Ferrús, Instituto Cajal, Madrid)

However, in neurological diseases, Jackson sees not only Spencer's principle of the dissolution of the nervous system in the form of neurological symptoms fulfilled, i.e. the failure of certain functions, but also his concept of "survival of the fittest" (this formulation actually comes from Spencer and not from Charles Darwin). With this term, Jackson refers to the fact that the resulting behaviour after a cerebral lesion is attributable to the behaviour of the remaining healthy ("fittest") brain structure. So if higher cerebral centres become functionless due to lesions, the resulting symptom is not only the absence of the higher function but the remaining function of the deeper brain structures, which thus take the position of the now highest centres. In this respect, even the most absurd thoughts and the most bizarre behaviour of neurological and psychiatric patients appear as a result of their remaining "healthy" brain structure. The misbehaviour, illusions and thought worlds of these patients thus represent their current and real mind.

Starting from his observations of neurological patients, Jackson not only developed theories about the function and interaction of cerebral structures but also designed a topical concept of the nervous system. Jackson assumes a vertical stratification of the brain into three levels. It is historically interesting in this context that Spencer also describes a development of the animal nervous system over three ascending levels in the *Principles of Psychology*. According to Jackson's idea, the lowest centres represent only a small part of the body, such as the anterior horns of the spinal cord or the eye muscle nuclei in the brainstem area. The middle centres re-represent the representations of the lower centres. Jackson calls these middle centres areas of double compound coordination. The highest centres finally re-represent the representations of the middle centres, thus becoming re-re-representative centres. In addition, these centres are the structures that first represent the entire body. Analogous to the multitude of lower and middle centres, Jackson also postulates several higher centres, so that the entire cortex cannot be regarded as a uniform higher centre. Each of these numerous higher centres rather represents the entire organism in different ways (Jackson 1931).

In line with this holographic view of the central nervous system, Jackson compared the brain to a "polyp whose individual elements have the same functions". However, the theory of the evolutionary structure of the brain also contains another aspect, namely that each element of an evolutionary level of the brain has a unique weighting for the function of a certain body part.

The analysis of the effects of neurological diseases on the concept of a hierarchical brain structure he postulated, led Jackson also to a theory of functional remission or neural plasticity. Based on his observation that a lesion in the corpus striatum resulted in a partial paralysis of the entire extremity and not a complete paralysis of specific muscles, Jackson concludes, on the one hand, that the nervous system represents movements and not muscles, and, on the other hand, that a functional remission is possible because the still functioning brain areas contain some representations of the damaged functions (York and Steinberg 1995).

This aspect of the different weighting of functions on various cerebral levels thus explains not only the clinical phenomenon of the recovery of a loss of function or rehabilitation after a specific brain lesion but also the phenomenon that a focal epileptic seizure does not immediately involve the entire body in the convulsions. The principle of compensating for a loss of function by still existing functioning elements in the nervous system applies to all evolutionary levels of the brain. Since higher centres have more complex and networked representations than lower brain areas, a lesion in the higher centre naturally leads to a lesser loss of function than in the lower centre. Although Jackson mainly refers to the motor system to illustrate these principles, he nevertheless assumed that the principle of compensation applies to all neural systems of the brain. In fact, modern imaging techniques suggest that patients with unilateral striatocapsular brain lesions, after appropriate neurorehabilitation, show activations in the ipsilateral striatum, in the primary motor cortex and the insula (this would correspond to Jackson's middle centre), but also in both frontal and parietal cortex areas (corresponding to the highest centres, Weiller et al. 1992). This recruitment of non-damaged areas at different levels during movement can thus be seen as empirical evidence of the ordinal representation in the central nervous system and the principle of compensation postulated by Jackson (York and Steinberg 1995).

Jackson began to apply his principle of evolution and dissolution to psychiatric diseases (Dewhurst 1982). From this point of view, a patient's hallucination appears as the "best" performance of lower brain regions, as a result of a disturbance of the higher brain structures. In the case of a patient who perceives a cat instead of a black hat—in the sense of an illusionary misidentification—the functioning lower brain regions now try to make the "best" of the incoming information and generate a corresponding perception. The patient's inability to perceive the object as a hat can be seen in this context as a negative element of his symptom (corresponding to the loss of function of higher brain areas) while the misidentification of the hat as a cat is considered a positive element (as an expression of the compensatory activity of lower structures).

Not only observations of patients with consciousness disorders but also his interest in philosophy since his youth prompted Jackson to engage with questions of consciousness and the emergence of the mind. In this context, he assumed that the mind, or the "self", is a manifestation of brain function but at the same time warned against equating mental states with neural states—thus equating mind with brain. In Jackson's conception, one state conditions the other, which is why we can speak of a psycho-physical parallelism. From a phylogenetic perspective, the emergence of the mind or the self does not require a new form of neural function that would have to be newly implemented in the brain; rather, the self is a manifestation or the result of a more complex coordination that is not yet possible in deeper brain regions. The special thing about this coordination centre is that there is now a new system of "unification" of the entire organism, whereby the entire organism is also adapted to the environment again. However, the emergence of the mind does require the evolution of new anatomical structures, which Jackson believes can be found in the prefrontal cortex. The "self" is not to be imagined as "residing" in the prefrontal cortex, but this new structure only enables a more complex coordination in this "sensomotor machine" postulated by Jackson. The principle of this "more complex coordination" can be most vividly illustrated by the process of becoming conscious. Jackson distinguishes between object and subject consciousness, whose anatomical substrates represent the body in its entirety. The whole brain not only represents the entire body, but it also represents every small unit of the entirety, albeit different aspects of the entirety from neighbouring units. This conception of brain organisation coincides with the principle of a hologram. In contrast to the units responsible for object consciousness, the units for subject consciousness are more organised and their functioning is more regulated or automated. Jackson illustrates his concept of consciousness using the sensation of a needle prick, where it seems evident to him that the conscious perception of the needle prick cannot be reduced solely to the concept of a "sensation". Just like the perception of a red light, the needle prick cannot be viewed as an isolated sensory phenomenon, but rather it represents a modification of states of consciousness. The needle prick induces impulses that reach the highest centres—which represent the entire organism—via afferent nerve fibres, in order to induce there a "increase in activity". The vivid, conscious perception that goes along with this activation, according to Jackson's theory, is synonymous with the "object consciousness" that is as it were "pushed out" from a pre-existing "subject consciousness". The anatomical substrates of subject and object consciousness appear in this conception as windowed monads, each of which—in line with Jackson's concept of motor representation—represents a slightly different aspect of the entire organism. The impulse induced by the needle prick subsequently finds its own and most suitable monad, which, according to Jackson's theory, represents more than just the sensation and localisation of the prick. The needle prick makes us more aware of our entire situation. Jackson compares this becoming aware with the image of awakening from a "mental absence" or from a daydream.

While Jackson admits that his ideas about mind and consciousness are speculative, they still correlate with his hierarchical concepts of the nervous system and form the foundation for current theories of consciousness research. The validity of the Jacksonian doctrine and the hierarchical organisation of the brain are confirmed by modern neurological research. In this context, hierarchical organisational principles could not only be found in the motor (Swash 1989) and sensory system (Vallbo 1989) but also in complex structures such as the visual system (Kennard 1989a) and the oculomotor system (Kennard 1989b). Even for the various aspects of memory performance, there are indications of a hierarchical functioning (Baddeley 1989).

2.2 Hierarchical Phenomena in Neurology

The morphological and functional similarities between the primate brain and the human brain suggest the presence of common phylogenetic roots. However, as the examination of these evolutionary residues is neither possible through natural science nor psychological methods, one is dependent on comparative morphology and behavioural research.

Instinctive movements are considered in the animal kingdom as innate behaviours, which just like morphological features belong to the characteristics of a certain species. In humans, such innate phenomena in pure form are almost exclusively found in newborns or infants, as early ontogenesis partially represents the evolutionary history. In adult humans, residues of instinctive actions are still found in rudimentary form, such as in the form of bowing or hat tipping as greeting gestures, which can be attributed to the instinctive submissive behaviour of many mammals. The sequence of instinctive actions observed by Konrad Lorenz in animals (the appetitive behaviour, the response of the innate triggering mechanism and the execution of the final action) is, however, hardly preserved in humans (Lorenz 1937). During ontogenesis, instinctive actions are increasingly integrated into more complex reflexive and voluntary movements due to the increasing maturation of the central nervous system. The second possibility to observe instinctive movements in humans is offered—according to the Jacksonian principle of dissolution—by conditions that lead to a cerebral lesion or disease processes that lead to a degradation of the brain. Depending on the degree of brain dysfunction, these instinctive movements occur reflexively or automatically. In the induced dissolution of brain function, phylogenetic and ontogenetic developmental stages are thus exposed or repeated. However, this recapitulation does not always proceed in a mirror image, as a reorganisation of the functional circuits takes place at each stage of degradation.

Only with the advent of comparative behavioural research were these instinctive movements in humans, which in clinical neurology were previously only considered as etiologically unclear symptoms, made explainable in terms of evolutionary history and accessible to systematic investigation. In modern neurology, these phenomena, also referred to as "primitive reflexes", are primarily of localising significance (Schott and Rossor 2002). Phenomenologically, in addition to oral instinctive movements, movements are also distinguished that can be attributed to grasping and holding functions, as well as disinhibitions of sexual behaviour, affective behaviour and expressive movements.

Oral Instinctive Movements

As part of the brain development of the infant, there is a chronological unfolding of predetermined cerebral functions. The development of oral instinctive movements also proceeds within the framework of ontogenesis from automatic processes to reflective—and thus environmentally related—behaviours. The so-called idle sucking can be named as the earliest oral instinctive movement in ontogenesis, which can be considered part of "idle breast searching" (Peiper and Thomas 1952). The oral automatism of "idle sucking" is possibly already associated intrauterine with the foetal swallowing of amniotic fluid and can later also be observed partly into the first years of life, especially in sleeping infants. An associated innate behaviour is the head swaying (Pilleri 1971) associated with breast searching (Prechtl and Schleidt 1950).

Oral instinctive movements have been observed in patients with various lesions or degenerative processes of the brain, according to the principle of dissolution of the nervous system. The characteristics of the movement patterns also correlate with the extent of the cerebral dysfunction, so that automatic movements occur primarily in the most pronounced and reflexive movements in lesser lesions. Oral automatisms, which in appearance correspond to the "idle sucking" of the infant, are described in patients with the highest degree of cerebral functional loss, the state of decerebration (Poeck and Hubach 1963). Head swaying movements, which in the infant represent part of the "idle breast searching", were observed in a patient with degenerative cerebral degradation process, i.e. Morbus Pick (Pilleri 1960a).

The instinctive movements of "oral grasping" represent, compared to the spontaneous oral automatisms, already more differentiated innate movement coordinations, which occur in infants, primates and other mammals, as well as in patients with certain brain lesions. Oral grasping corresponds to the tendency to hold a biting object between the upper and lower jaw (Pilleri and Poeck 1964).

This phenomenon can already be observed in newborns and can be triggered tactilely by both proprioceptive stimuli (i.e. by touching the lips) and exteroceptive stimuli (e.g. touches of the perioral region, the ear or neck region). Characteristic of oral grasping is that the tonic holding of the object in the mouth is further intensified by pulling on the object, so this phenomenon is also referred to as the "bulldog reflex" (Pilleri and Poeck 1964).

Upon touching the mouth and cheek region, the "rooting" phenomenon may also occur, in which the corner of the mouth initially moves towards the side of the stimulus and subsequently leads to the mouth opening (Pilleri and Poeck 1964).

Optically triggered oral grasping, compared to the tactilely triggered reactions, represents a more differentiated instinctive movement, which only occurs in infants from about the fourth month of life (Pilleri and Poeck 1964).

Animal observations confirm that oral grasping movements in gorilla infants are initially elicited only tactile and from about the third month of life can also be triggered by optical stimuli (Lang and Schenkel 1960–61).

In topodiagnostic terms, instinctive movements of oral grasping have been described primarily in temporal, especially in mediobasal brain lesions (Pilleri 1961). Correlating with these findings is the fact that oral automatisms such as swallowing, smacking or chewing are specific symptoms of temporal lobe epilepsy. In addition, the same motor phenomena can also be triggered by electrical stimulation of the temporal lobe.

A lesion in the area of the cortical representations of the chewing-, mouth- or facial muscles (or the efferent cortico-pontine pathways) leads to a central paralysis of the corresponding muscles but not to the mentioned oral automatisms. These are thus not released by interruption of the voluntary motor pathways but are disinhibited by temporal lesions and thus clinically visible. Analogous to the chronological occurrence of the instinctive movements in the infant, there is a mirror-image reappearance of the oral movement patterns in slowly progressing brain lesions (such as in cerebral degradation processes), depending on the extent of the functional disorder. In the case of minor damage, only optically triggerable oral grasping is initially manifested, however with increasing cerebral disturbance tactilely triggerable oral grasping becomes manifest (Pilleri 1961). In the final stage of degradation processes, automatic movement patterns such as breast-seeking and sucking in idle reappear. The successive functional dissolution of the central nervous system at various hierarchical levels thus leads clinically comprehensibly to a (nearly mirror-image) recapitulation of both onto- and phylogenesis.

Instinctive Grasping and Holding

Even with the instinctive movements of grasping, the automatic movement patterns are considered the most primitive or undifferentiated motor sequences. The alternating up-and-down movements of both arms observable in infants in supine position—where the hands are simultaneously opened and closed—are considered in the literature as climbing automatisms (Prechtl 1953). Since these automatisms are not found in all infants, it is suspected that these phenomena in humans are already in phylogenetic decline. The phylogenetic significance of the climbing and grasping templates is to be found for the newborn in the original function of clinging to the mother's fur; these automatisms were also useful for the adult ancestors of humans for movement in the trees. Correspondingly, climbing automatisms can only be demonstrated in tree-climbing chimpanzees, orangutans and gibbons.

There are individual case reports of the occurrence of climbing automatisms in brain lesions in patients in the final stage of cerebral degradation processes (Pilleri 1960b).

Analogous to the oral instinctive movements, the instinctive grasping movements can also be triggered proprioceptively (by stretching the finger flexors) or exteroceptively (by touch stimuli on the palm of the hand). When the finger flexors are stretched, there is a reflexive closing of the fingers, which leads to a tonic holding of the object. These movement coordinations belong to the innate motor templates of the infant, where in addition to hand grasping reflexes, weaker foot grasping reflexes (or "podal" grasping reflexes) can also be observed (Wieser 1957a). Both reflexes appear in the same intensity only in premature births. The unequal expression of both grasping reflexes in the mature infant is phylogenetically explained by the transformation of the foot from a grasping tool to a walking tool. The podal grasping reflexes are thus, like the climbing automatisms in humans, in phylogenetic regression. In all tree-dwelling primates, the tonic proprioceptive grasping reflexes are correspondingly well developed (Pilleri & Poeck 1964).

In brain-damaged patients, the reappearance of the grasping reflexes can lead to such a firm hold of objects (Pilleri 1971) that the patient can sometimes be lifted from the surface (Wechsler et al. 1936).

The instinctive movement of the so-called "reaching out" is only triggered by optical activation of the grasping movement, therefore by moving towards or pulling away an object. This function already presupposes a (visual) interaction with the environment and is therefore only detectable in humans in the third to fourth month of life; at the same time, tactile instinctive movements are reduced triggerable. Observations on primates were also able to confirm this chronological principle of instinctive movements (Lang and Schenkel 1960–61). In the brain-damaged patient, the grasping can also be induced by a moving object, in which case a (turning-) movement can be triggered even before the object is grasped by the patient which may even involve the entire body (Wieser 1957b).

Locally, instinctive grasping movements occur with lesions of the precentral region of the frontal lobe, while foci in the temporal lobe are not associated with it (Fulton 1934). Similar to the oral templates, grasping movements are therefore not caused by lesions of the pyramidal tract but by interruption of other inhibitory (in the case of the grasping movements frontal) pathways. The same applies to the manifestation of the grasping reflexes, that onto- or phylogenesis is almost repeated in a mirror-like manner with progressing cerebral degradation. Minor brain damage leads to optically triggered grasping, while moderate lesions lead to tactilely triggerable grasping movements, and in the final stage of cerebral degradation processes, automatic grasping movements occur again.

Disinhibition of Sexual Behaviour

The previously described instinctive forms of movement of grasping, as well as oral grasping, are from an ethological point of view the end part of instinctive actions, namely the so-called final actions, or consummatory acts. If there is primarily a disturbance of the drive, this is expressed by a pathological course of the entire instinctive action, which can affect both sexual and affective behaviour.

Brain damage leading to a disinhibition of sexual drive behaviour is primarily located in the medial temporal lobe. The symptoms can range from a purely verbal disinhibition to an extremely increased sexual desire, which can sometimes be associated with persistent erections and excessive masturbation. In this context, it should also be noted that sexual phenomena such as erections, sexual arousal or masturbatory acts can be symptoms of a psychomotor epilepsy, whose seizure origin is also found in the medial temporal lobe. In addition to this location, increased drive behaviour in humans has also been observed with lesions of the hypothalamus, selective bilateral lesions of the hippocampus, as well as the amygdala. In the context of rabies, whose foci are primarily located in the hippocampus, it can sometimes lead to excessive sexual behaviour with nymphomania and satyriasis.

Masturbatory behaviour, sometimes also with excessive symptoms, is extensively documented from animal observations—especially in primates—but has also been described in ungulates, carnivores and even in some bird species.

Anal behaviour, i.e. the occupation with faeces, also belongs to the innate behavioural repertoire of the infant, which not only in psychoanalytic terms can be classified under the concept of instinctive behaviour. In the case of cerebral degradation processes, there can be a recurrence of actions, such as encopresis, smearing of faeces and even coprophagy, so that here too an inhibition of the aforementioned behaviour can be assumed during the course of ontogenesis. There are few empirical data on anal behaviours in primates, but phenomena such as coprophagy have already been described in rodents.

Disinhibition of Affective Behaviour

From the perspective of ethology, aggression in animals is an integral part of their social behaviour and thus also an essential factor in natural selection. In humans, this affective state has been subjected to significant inhibition and sublimation as part of cultural development. In this respect, phenomena of disinhibition of anger behaviour in humans represent a phylogenetic regression. Patients with such symptoms react to the slightest triggers with massive states of arousal in the form of aggressive actions up to physical violence. The outbursts of rage themselves are often described by the patients as "anger that arises against their will". This pathological rage behaviour differs from normal human anger outbursts in that it can be triggered unspecifically, is excessive in scale and uncontrolled and unstoppable in its course. The lack of fatigue is another characteristic of this pathological behaviour. The same criteria also apply to the previously discussed motor disinhibition phenomena.

The cerebral lesions that lead to disinhibition of rage behaviour in humans are primarily located in midline structures, especially in the septum region and in the hypothalamus. There are also reports of aggression symptoms in cases of bilateral medial temporal lobe damage (Bingley 1958). Lesions or electrical stimulation of the same cerebral regions in animals also lead to the same pathological rage reactions as in humans (Hess 1949). Extensive removals of the forebrain induce in animals the phenomenon of sham rage, in which only the motor and vegetative components of the rage reaction occur; the actual affective reaction—as far as can be measured in animal experiments—however, can no longer be observed (Cannon 1953). Symptoms of sham rage have also been described in patients with midline cerebral lesions that led to an interruption of pathways between the forebrain and the brainstem (Wortis and Maurer 1942).

Conditions associated with a reduction of affect and drive in the form of particular gentleness, manageability and spontaneity have also been described in humans. In animal experiments, such a reduction of affect or "placidity" could be induced after bilateral medial temporal lobe resection and bilateral amygdala lesion (Schreiner and Kling 1956). In humans, there are case reports of this affective disorder associated with bilateral lesions of the cingulate gyrus and the medial temporal lobe structures. Since this disorder leads not only to a change in affect but also in drive, it cannot be assumed to be an analogous pathology to the disinhibited rage reactions. Furthermore, no homologous behaviours from phylogeny are known in this regard. In terms of brain topography, it is interesting to note that those brain areas in the limbic system, whose lesion disinhibits aggressive behaviour, cover a significantly larger area than those, whose damage results in a reduction of affect with extreme gentleness. These significant differences on an anatomical level underline the biological significance of aggression in social behaviour.

Disinhibition of Expressive Behaviour

For human social behaviour, in addition to language, the expressive movements of laughter and crying are also significant. While parts of this expressive behaviour are innate, other components mature during ontogenesis to eventually form a permanent connection with the affective system. The fact that even blind-born children smile supports the assumption that laughter and crying are—at least partially—instinctive actions. Phenomenologically, the so-called "scream crying" can initially be observed in newborns, which initially occurs without tears. Crying without screaming only begins after a few months. Smiling or laughing appears as a later and thus more mature manifestation of expressive movements.

During ontogenesis, the expressive movements initially seem to be primarily signs of a vegetative reaction, while they later increasingly become an expression of affective experience. Characteristic of differentiated expressive movements is also the response to external stimulation. The change in the appearance of these instinctive movements suggests that the neural substrate of these templates is located at various levels of the brain and matures at different intervals. The vital facial reactions of displeasure seem to be organised at the brainstem level at the beginning of ontogenesis, while the reactions of smiling or laughing—which primarily aim at increased social interaction—are represented in higher brain regions.

In primates, especially in chimpanzees, expressions comparable to laughter and crying have been described, both in response to external stimuli and as spontaneous behaviour (Pilleri 1971).

These animals thus react in situations that are affectively comprehensible for humans, with laughter or crying. However, the crying of monkeys always remains tearless, just like the scream crying of human infants. The development of expressive movements in primates follows the same sequence as in human infants: from undifferentiated crying, to smiling, to laughing.

The instinctive expressive movements in adult humans are only visible as disinhibition phenomena in the sense of the so-called pathological laughter and crying in the case of corresponding brain lesions (Poeck and Pilleri 1963). Clinically, the affected patients respond to unspecific external stimuli with sudden and uncontrolled laughter or crying, with corresponding acoustic, respiratory and vegetative appearances. The patients also experience the symptoms as an expression not conforming to the current affective state. In some cases, there is also a transition from laughter to crying, and vice versa (Pilleri and Poeck 1964).

Pathological laughter can occasionally also occur as a symptom of an epileptic seizure, when only "laughing fits" occur, it is referred to as "gelastic epilepsy".

The Snout or Shovel Phenomenon

An instinctual movement that is not clearly categorisable is the snout or shovel phenomenon already described by Darwin ("pursing of the lips", Darwin 1872). In this appearance, there occurs a spontaneous or by perioral touch induced snout-like protrusion of the lips, without these being tightly closed or even associated with sucking movements. The phenomenon occurs spontaneously in newborns and is observed into the first years of life; occasionally, it can be seen in healthy adults in a state of increased concentration. In primates, this movement occurs in various states, both in irritation, disgust, and phases of increased attention (Pilleri 1971).

As a sign of disinhibition, the snout phenomenon is observed in diffuse cerebral degradation processes (Pilleri 1971).

Most authors agree with Darwin's view that this sign belongs to the instinctive expressive movements, and not to the (oral) instinctive movements of food intake.

Bibliography

Baddeley A (1989) Hierarchies and human memory. In: Kennard C, Swash M (Eds) Hierarchies in neurology. A reappraisal of a Jacksonian concept. Springer-Verlag, London, New York, pp 49–54.

Bingley T (1958) Mental symptoms in temporal lobe epilepsy and temporal lobe gliomas. Acta Psychiat. Neurol. Scand. 33, Suppl. 120.

Cannon WB (1953) Bodily changes in pain, hunger, fear and rage. 2nd ed. Borton, Boston.

Darwin CR (1872). The expression of the emotions in man and animals. John Murray, London.

Dewhurst K (1982) Hughlings Jackson on psychiatry. Sanford, Oxford.

Fulton JF (1934) Forced grasping and groping in relation to the syndrome of the premotor area. Arch. Neurol. Psychiat. 31: 221–235.

Hess WR (1949) Das Zwischenhirn. Syndrome, Lokalisationen, Funktionen. Schwabe, Basel.

Jackson JH (1931) Taylor J. (Ed) Selected writings of John Hughlings Jackson. Staples Press, London.

Kennard C (1989a) Hierarchies and the visual system. In: Kennard C, Swash M (Eds) Hierarchies in neurology. A reappraisal of a Jacksonian concept. Springer-Verlag, London, New York, pp 87–92.

Kennard C (1989b) Hierarchical aspects of eye movement disorders. In: Kennard C, Swash M (Eds) Hierarchies in neurology. A reappraisal of a Jacksonian concept. Springer–Verlag, London, New York, pp 151–158.

Lang EM, Schenkel R (1960–61) Goma, das Basler Gorillakind. Documenta Geigy, Basel.

Lorenz K (1937) Bildung des Instinktbegriffes. Naturwissenschaften 25: 289–300.

Peiper A, Thomas H (1952) Leerlaufendes Brustsuchen. Mschr Kinderheilk. 101: 303–307.

Pilleri G (1960a) Kopfpendeln („leerlaufendes Brustsuchen“) bei einem Fall von Pick’scher Krankheit. Arch. Psychiat. Nervenkr. 200: 603–611.

Pilleri G (1960b) Über das Auftreten von Kletterbewegungen im Endstadium eines Falles von Morbus Alzheimer. Arch. Psychiat. Nervenkr. 200: 455–461.

Pilleri G (1961) Über das chronologische Auftreten von motorischen Schablonen des Oralsinnes, deren ontogenetische Bedeutung und klinisch-anatomische Zusammen- hänge bei atrophisierenden Hirnerkrankungen. Schweiz. Arch. Neurol. Neurochir. Psych. 88: 273–298.

Pilleri G (1971) Instinktbewegungen des Menschen in biologischer und neuropathologischer Sicht. Akt. Fragen Psychiat. Neurol. 11: 1–37.

Pilleri G, Poeck K (1964) Arterhaltende und soziale Instinktbewegungen als neurologische Symptome beim Menschen. Psychiat. Neurol. 147: 193–238.

Poeck K, Hubach H (1963) Rhythmische orale Automatismen bei Dezerebrationszuständen. Deutsch. Z. Nervenheilk. 185: 37–52.

Poeck K, Pilleri G (1963) Pathologisches Lachen und Weinen. Schweiz. Arch. Neurol. Neurochir. 92: 323–370.

Prechtl HFR, Schleidt WM (1950) Auslösende und steuernde Mechanismen des Saugaktes. Z. vgl. Physiol. 32: 257–262.

Prechtl HFR (1953) Über die Koppelung von Saugen und Greifen beim Säugling. Naturwissenschaften 40: 347.

Ramón y Cajal S (1892) El nuevo concepto de la histologia de los centros nervi-osos. III. Corteza gris del cerebro. Rev. Cien. Med. Barcelona 18: 457–476.

Ramón y Cajal S (1894) Les nouvelles idées sur la structure du système nerveux chez l'homme et chez les vertébres. Reinwald, Paris.

Ramón y Cajal S (1909–1911) Histologie du système nerveux de l'homme et des verté- brés. 2 Vol. Maloine, Paris.

Schott JM, Rossor MN (2002) The grasp and other primitive reflexes. J Neurol Neurosurg Psychiatry 74: 558–560.

Schreiner L, Kling A (1956). Rhinencephalon and behaviour. Amer. J. Physiol. 184: 486–490.

Spencer H (1855) The principles of psychology. London: Longman, Brown, Green, and Longmans.

Swash M (1989) Order and disorder in the motor system. In: Kennard C, Swash M (Eds) Hierarchies in neurology. A reappraisal of a Jacksonian concept. Springer-Verlag, London, New York, pp 113–122.

Taylor J. (ed) (1931/32) Selected writings of John Hughlings Jackson, vols 1 and 2. Hodder and Stoughton, London. Reprinted (1958) Basic Books, New York.

Vallbo AB (1989) Single fibre microneurography and sensation. In: Kennard C, Swash M (Eds) Hierarchies in neurology. A reappraisal of a Jacksonian concept. Springer-Verlag, London, New York, pp 93–109.

Wechsler JS, Bieber J, Balser GH (1936) Postural reflexes in patients with lesions of the frontal lobe. Arch. Neurol. Psychiat. 35: 1208–1215.

Weiller C, Chollet F, Friston KJ, et al. (1992) Functional reorganization of the brain in recovery from striatocapsular infarction in man. Ann Neurol 31: 463–472.

Wieser S (1957a) Pathologie und Physiologie des Greifens. Fortschr. Neurol. Psychiat. 25: 317–341.

Wieser S (1957b) Schlüsselreize raumorientierender Zuwendereaktionen. Arch. Psychiat. Nervenkr. 195: 373–382.

Wortis H, Maurer WA (1942) Sham rage in man. Amer. J. Psychiat. 98: 638.

York GK, Steinberg DA (1995) Hughlings Jackson's theory of recovery. Neurology 45: 834–838.

CHAPTER 3

Hierarchical Concepts in Neuroethology

3.1 An Evolutionary Approach to Brain Function

The first theories regarding a hierarchical organisation of behaviour were already put forward by behavioural research in the middle of the last century. Tinbergen, in his concept of the "instinct hierarchy", assumes that there are superordinate and subordinate drives. He defines instinct as a "hierarchically organised, nervous mechanism that responds to certain warning, triggering and directing impulses, both internal and external, and answers them with well-coordinated, life- and species-preserving movements". According to these views, the hierarchical organisation of behaviour is based on a corresponding hierarchical structure of functional instances of the brain (Tinbergen 1951). Weiss already postulated a central nervous hierarchy correlating with behaviour in 1941, which consists of several integration centres.

This concept is very reminiscent in its form of the hierarchical-neurological concept of Hughlings Jackson (Weiss 1941). A topically differentiated hierarchy model was presented by Baerends (Fig. 3.1), which emphasises the networking of the systems with each other and assumes that hierarchically subordinate centres are often controlled by several superordinate ones (Baerends 1956).

G. Wiest, *Hierarchies in the Brain, Mind and Behaviour*, https://doi.org/10.1007/978-3-031-78030-1_3

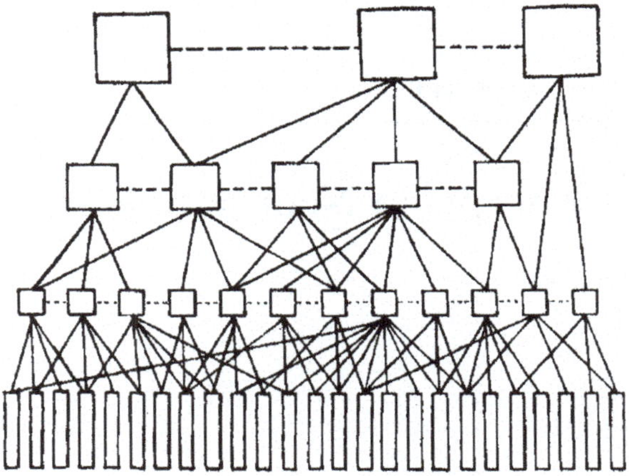

Fig. 3.1 Hierarchical organisational scheme of behaviour according to Baerends, Instinkt höherer Ordnung = Higher order instinct, Instinkt niedriger Ordnung = Low-order instinct, Erbkoordination = fixed action pattern, Muskelbewegung = Muscle movement (Baerends 1956). (With kind permission from Prof. Irenäus Eibl-Eibesfeldt)

The empirical confirmation of the hierarchical organisation of the brain and behaviour at the neural level was only achieved through the neurobiological data of neuroethology.

A theory of brain function from an evolutionary or hierarchical perspective requires both a reductionist and a holistic approach. Converging data from comparative anatomy, palaeontology as well as ontogenesis prove that during the phylogeny of humans, there was primarily a growth of the forebrain (prosencephalon), which is composed of the endbrain or cerebrum (telencephalon) and the interbrain (diencephalon) (Fig. 3.2).

At the same time, however, similarities of three neural associations, which represent the phylogenetic heritage of reptiles, lower and higher developed mammals, were maintained, so that from an evolutionary perspective, one differentiates between a so-called reptile brain, paleomammalian brain and neo-mammalian brain. According to the theories of evolutionary neuroethology (MacLean 1990), these associations, although they differ significantly in their structure and neurochemistry and are aeons apart in evolutionary terms, form a unit, corresponding to a conglomerate of three brains. According to Paul MacLean, the human brain from an evolutionary perspective is thus a "Triune Brain" (Fig. 3.3).

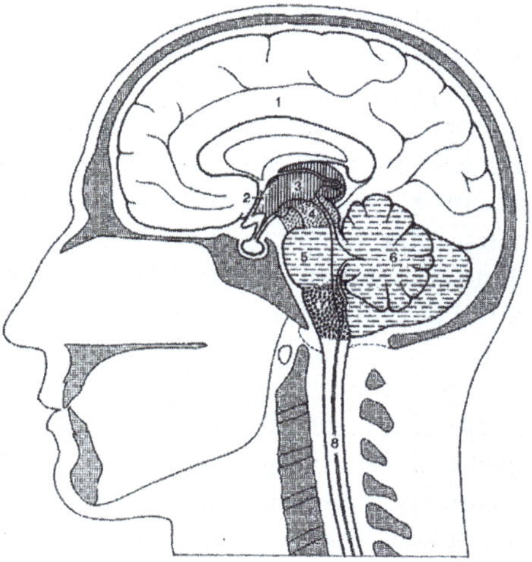

Fig. 3.2 Neuroanatomy of the central nervous system. *The Human Central Nervous System* (Nieuwenhuys et al. 2007). (Copyright: Springer) 1. Telencephalon 2. Telencephalon impar 3. Diencephalon 4. Mesencephalon 5. Pons 6. Cerebellum 7. Medulla oblongata 8. Medulla spinalis

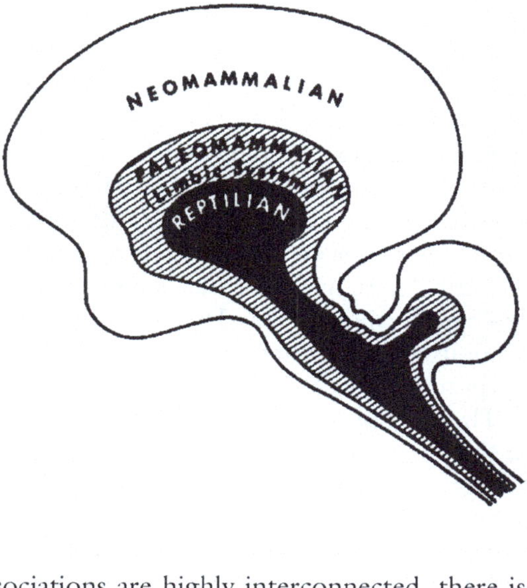

Fig. 3.3 The triune brain model by Paul MacLean (MacLean 1990) (Copyright: Springer)

Despite the fact that the associations are highly interconnected, there is also evidence for independent functioning. A peculiarity in this regard arises from the fact that the two older neural associations structurally and functionally have no possibility of exchanging verbal information.

The three brain formations of the "Triune Brain" can be presented as three evolutionary neural systems that are interconnected. However, each system possesses its own form of intelligence, its own subjectivity and memory, as well as its own feelings of time and space. The networking and interactions of the individual associations with each other also mean that the entirety of the brain, from a functional perspective, represents more than just the sum of its individual parts. The topographical structure of the "Triune Brain" should not be misinterpreted as being a purely chronological layering of neural associations, similar to geological layers of rock formations. Equally erroneous is the assumption that the phylogenetic correlate of the reptilian brain of mammals corresponds to that of currently existing reptiles; this would completely disregard the evolutionary development of these associations.

In current neurobiological basic research, there is no doubt that the data obtained at the cellular and molecular level in animal models can also be applied to the biology of the human body. It is all the more surprising, in this respect, the doubts of some scientists when it comes to comparing or homologising animal and human behaviours. Evolutionary neuroethology not only aims to combine the methods of comparative behavioural research with neurobiology, but it also enables a new form of epistemology. In contrast to the field of research in philosophy known as epistemology, which deals with the possibility of knowledge of the subject, evolutionary neuroethology deals with the relationship of the subjective self to the inner world and to external reality. Currently, only psychoanalysis, which is often criticised from a scientific perspective for its lack of scientific rigour, meets this research requirement. Paul MacLean proposed the term "epistemics" for such a research direction, which, including psychological, ethological and neuroscientific disciplines, seeks to explain the relationship of the self to inner experience and external reality—analogous to epistemology (MacLean 1975a).

The description of behaviours, which are brain topographically assigned to the phylogenetically older neural associations, requires its own terminology of non-verbal behaviour, primarily to ensure a homologisation between human and animal ("non-verbal") behaviour patterns. Comparative behavioural research was the first scientific discipline to deal with the homologisation of non-verbal behaviour in humans and animals. The pioneer of evolutionary neuroethology, Paul MacLean, proposed the term "prosematic behaviour" to describe human and animal non-verbal behaviour, which also includes non-verbal communication via physical, acoustic or chemical signals (MacLean 1977). The interest of evolutionary neuroethology lies in the identification of neural associations in the forebrain that are associated with prosematic behaviours (which in turn express so-called paleopsychic processes). Paul MacLean introduced the terms "protomentation" and "emotional mentation" to describe so-called paleopsychic processes. The term protomentation refers to psychic functions that are involved in the regulation of basic behavioural routines and certain patterns of non-verbal communication. The term emotomentation describes, on the other hand, a mental activity that regulates or influences behaviour based on the subjective perception of emotions. In analogy to these non-verbal functions, "ratiomentation" would thus correspond to the rational, self-conscious activity of the human neocortex, which alone has the possibility of linguistic communication.

According to MacLean, the brain formations of the forebrain—according to their evolutionary structure and their different behavioural functions—are divided into three neural associations: The reptilian brain or the proto-reptilian formation (R-complex); the mammalian brain, which is associated with the limbic system; as well as the neo-mammalian brain, which corresponds to the neocortex.

3.2 The Reptilian Brain (R-Complex)

This oldest part of the forebrain developed about 300 million years ago as a specific brain formation of the reptile-like ancestors of mammals, the therapsids.

The neural associations of the R-complex show remarkable structural constancy in the course of evolutionary development, which is why neuroanatomists of the nineteenth century already attributed an important functional significance to them. Morphologically, the reptilian brain is a defined structure of grey matter in the basal parts of the forebrain. In primates, these structures can be anatomically assigned to the olfactostriatum, corpus striatum (which is formed from the caudate nucleus and putamen), the globus pallidus, the substantia innominata and the Meynert basal nucleus. In neurological terminology, these associations are referred to or summarised as part of the basal ganglia, and in humans, they represent the central units of the so-called extrapyramidal-motor system (Fig. 3.4). However, since the reptilian brain does not include all brain structures of the basal ganglia (e.g. the claustrum), MacLean proposed the term "striatal complex" or "R-complex" to distinguish it from these (MacLean 1973).

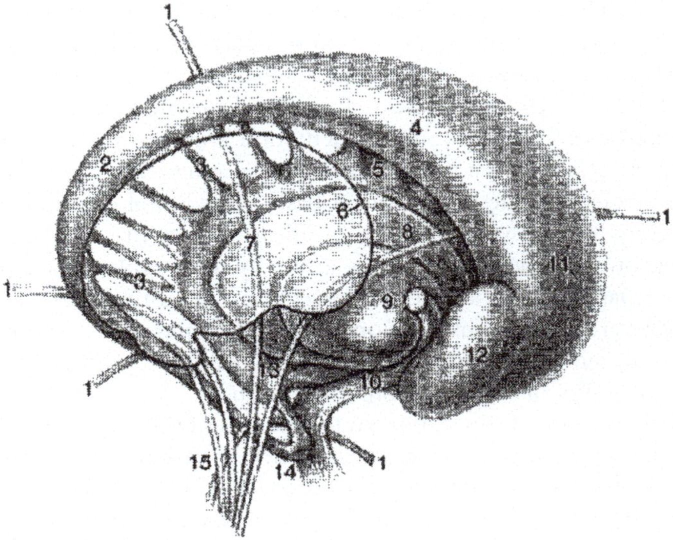

Fig. 3.4 The structures of the basal ganglia. *The Human Central Nervous System* (Nieuwenhuys et al. 2007). (Copyright: Springer) 1. Corona radiata 2. Tail of the caudate nucleus 3. Caudatolenticular grey bridges 4. Body of the caudate nucleus 5. Putamen 6. Thalamus 7. Internal capsule 8. Lateral globus pallidus 9. Medial globus pallidus 10. Anterior commissure 11. Head of the caudate nucleus 12. Nucleus accumbens 13. Peduncle of the lentiform nucleus 14. Connection of the tail with 13 15. Cerebral peduncle

The extinction of the therapsids, the presumed ancestors of mammals, prevents a direct neuroethological investigation and homologisation of R-complex-specific behaviours. These reptile-like therapsids inhabited the then mega-continent Pangea until about 190 million years ago and, as ancestors of mammals (this assumption is based on palaeobiological data), were equipped with a kind of "proto-striatal complex" that went hand in hand with the evolution of basal behavioural routines (Fig. 3.5).

One possibility of comparative neuroethological investigation of R-complex-associated behaviours arises in the study of the behaviour of lizards, as these, of all living reptiles, show by far the highest similarity to the reptile-like ancestors of mammals.

Indeed, comparative-anatomical studies on embryos of lizards, birds and mammals suggest that certain homologies in the structure of the

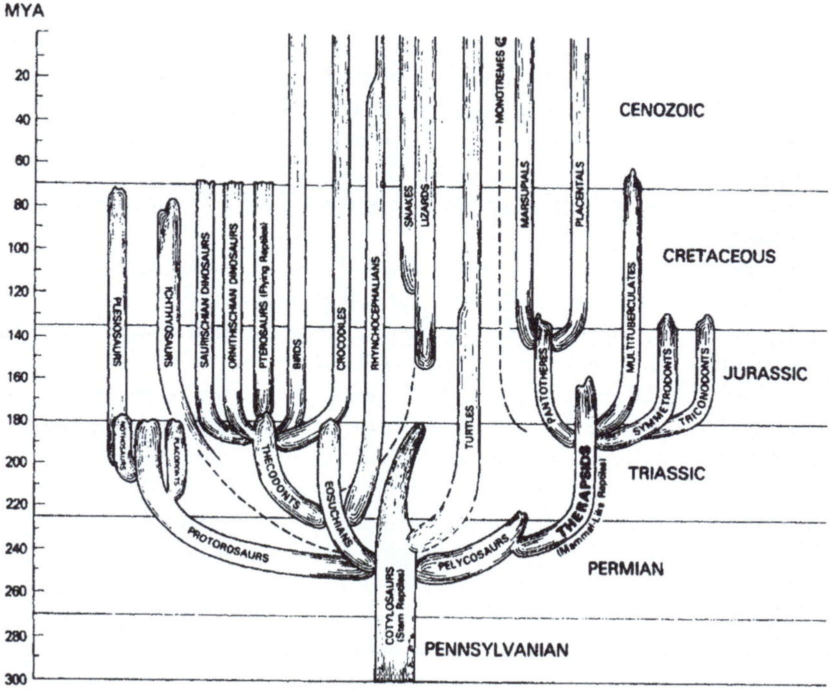

Fig. 3.5 Representation of the phylogeny of reptiles and mammals (MacLean 1990). (Copyright: Springer)

developing striatal complex can be demonstrated (Kallen 1951). Although Haeckel's biogenetic principle, namely that ontogeny recapitulates phylogeny, cannot be adopted without reservation, contemporary discussions of this principle confirm astonishing agreements (Gould 1977). Recent comparative-anatomical data clearly indicate that the structure of the striatal complex—starting from a rudimentary approach in amphibians—shows astonishing constancy in reptiles and mammals (Striedter 2005), in the sense of a fundamental scheme (Marín et al. 1998).

Neuroanatomical data also prove that there has been a significant increase in the afferents and efferents of the striatal complex over the course of evolution.

The corpus striatum of mammals is reached by afferent pathways from the reticular system, the substantia nigra, the intralaminar nuclei, the cingulate cortex and the amygdala, as well as extensive areas of the neocortex. The efferents of the corpus striatum have connections with the globus pallidus and the substantia nigra. Efferents from the globus pallidus, on the other hand, reach ventral and medial parts of the thalamus, the subthalamic region and the ventral tegmentum (Panksepp 1998). Apart from the connections of the olfactory organ with the so-called olfactostriatum, no additional afferents from other sensory systems seem to exist to reach the reptilian brain.

The discussion about the function of the striatal complex continues despite a wealth of animal experimental and clinical data.

The debate is controversial. Above all, the traditional view of neurology, that the corpus striatum is exclusively a centre of the extrapyramidal-motor system, led primarily in clinical research to a focus on the motor functions and to a neglect of cognitive or behavioural aspects. The animal experimental and clinical results, obtained through lesion studies, cerebral stimulation as well as recording neural activity, do not provide a uniform picture of a specific motor function of these structures. During electrical stimulation of the striatal complex of reptiles and birds, no specific movement patterns could be induced (Goldby and Gamble 1957; Anschel 1977). Electrical stimulation of the globus pallidus in humans induced only turning movements of the body to the opposite side (Hassler 1961), in the striatum, however, it induced disturbances of mood and affect, as well as attention (Van Buren et al. 1966). Electrophysiological studies (single unit recordings) could also demonstrate that the majority of striatal neurons have no association with specific movements (Rolls et al. 1979). Since the corpus striatum could not be assigned either clearly motor functions in the sense of a "motor centre" or sensory functions, it was initially considered an unspecific "integration centre", whose importance was yet to be determined (Albe-Fessard and Buser 1979). Current neurobiological data point to a significance of the basal ganglia for learning and memory processes. According to these views, the striatum and other core areas of the basal ganglia receive information signals about the current state of motor or mental events and are thus with other formations part of a comprehensive adaptive system that optimises the execution of actions (Graybiel 2008).

One aspect that has been neglected in the previous research on the function of the striatum and brain function in general was the fact that symptoms or phenomena induced by cerebral lesions, stimulations or recordings do not only manifest as so-called positive or observable phenomena but that, according to Hughlings Jackson, "negative symptoms" can also occur, which can escape experimental data collection. In contrast to traditional neurobiological research on animals, which often uses very standardised experimental paradigms and methods, comparative neuroethology aims to draw conclusions about the function of certain brain structures by observing natural animal behaviours.

Recent neuroscientific research approaches or data that associate the striatum with the formation of neural representations of habitual actions, however, already show a remarkable agreement with the research results obtained on lizards, mammals and humans in comparative neuroethology on the function of the striatal complex.

The R-complex and the Behaviour of Lizards

Evolutionary neuroethology attempts to describe homologies of behavioural patterns in phylogenesis using the method of comparative behavioural research, and to gain insight into the function of certain cerebral structures. In the case of the striatal or R-complex, difficulties arise insofar as the first and most primitive form of the reptilian brain of mammals is no longer accessible for research due to the extinction of the therapsids. Although none of the currently existing reptiles share the same lineage with the reptile-like ancestors of mammals, lizards offer themselves for neuroethological examination of the R-complex due to their anatomical similarities. The observation of the natural behaviour of lizards, including their communication behaviour with each other, allows—in contrast to pure neurobiological research, which mainly investigates learning processes or fear and avoidance behaviour using standardised paradigms—a capture of R-complex-associated prosematic behaviours (MacLean 1977).

Studies on lizards prove that under laboratory conditions they show the same behaviours as in the wild. Greenberg et al. (1976) was able to demonstrate that the behaviour of lizards throughout the day is characterised by certain routine processes (daily master routines).

These routines include (1) cautiously emerging from a hiding place, (2) a period of sunning, (3) followed by defecation at an assigned place, (4) morning food intake within the territory, (5) a period of afternoon rest, (6) afternoon food intake in the terrain, (7) afternoon sunning, (8) return to the hiding place. In addition to this stereotypical structure of the daily routine, lizards also show four specific behavioural patterns or "displays":

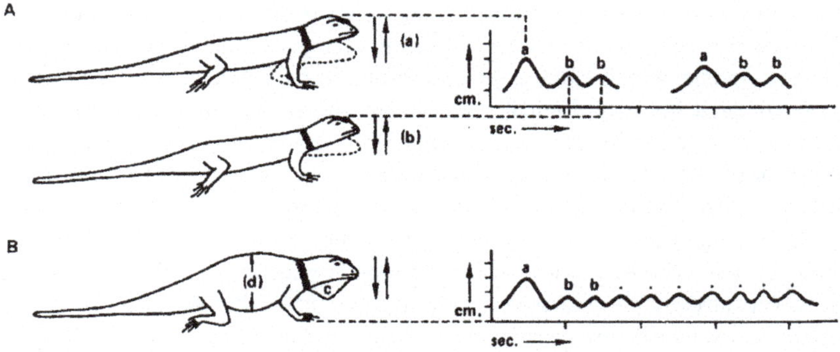

Fig. 3.6 Characteristics of recognition actions (A) and threat or attack actions (B) of lizards (MacLean 1990). (Copyright: Springer)

(1) Recognition actions: This behavioural pattern consists of a short flexion and extension movement of the front extremities, followed by two head nods (Fig. 3.6A). These actions occur, on the one hand, when two or more lizards meet in the sense of a greeting ritual (here there are certain homologies to human greeting gestures); on the other hand, they are also used as expressive behaviour in front of territorial intruders and at the beginning of courtship behaviour. (2) Threat or attack actions: These occur on contact with intruders and lead to an enlargement of the neck crease and increase in body height to a threateningly appearing expansion of the body circumference (Fig. 3.6B). Subsequently, it can lead to the positioning of two opponents side by side (Fig. 3.7A), upon which the actual attack takes place. (3) Courtship behaviours: These are often initiated by the female, by signalling readiness to mate through characteristic tail movements (Fig. 3.7B). The female's readiness to mate also includes a lifting of the tail and the presentation of the rear part of the body. The subsequent behavioural patterns of the male often resemble recognition actions, followed by territorial actions and finally the act of copulation. (4) Submission or humility behaviours: These represent an important survival strategy in the behavioural repertoire of all animals. In lizards, the limbs are drawn to the body, followed by a lowering of the body and the head. Neuroethological studies on larger lizards and monitor lizards confirmed that these perform the same behavioural routines and characteristic actions as part of their prosematic communication as lizards do (Harris 1964; Auffenberg 1981).

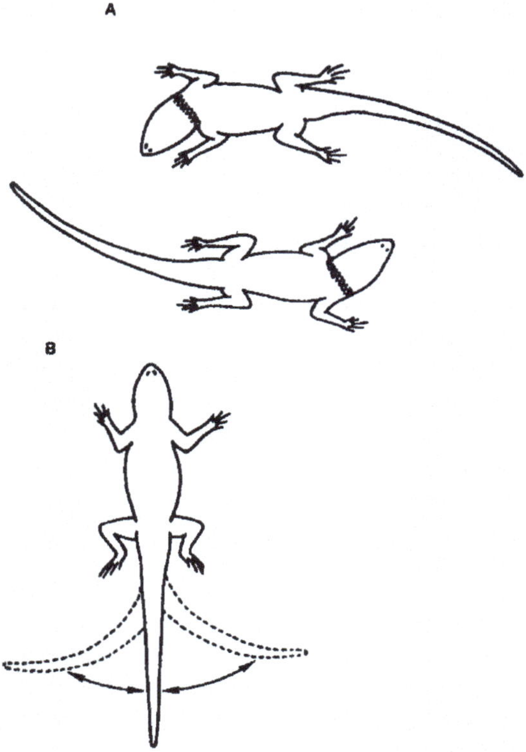

Fig. 3.7 Positioning of opponents side by side before an attack (A) and characteristic tail movements (B) as courtship behaviour in lizards (MacLean 1990). (Copyright: Springer)

The basic behaviours of lizards listed so far can be extended by some additional "special" forms of behaviour, which are at least rudimentarily found in these reptiles. The so-called grooming or cleaning behaviours are primarily evident in more highly developed animal species, but they can at least be observed in lizards, especially as part of the shedding process. Greenberg (1978) reported forms of social grooming behaviours in lizards, which were performed on conspecifics during shedding. A more common type of grooming behaviour, namely the rubbing of the mouth region after feeding, was confirmed by several authors (Auffenberg 1978).

The breeding behaviour gained importance during the course of evolution only with the emergence of mammals. Lizards show no signs of breeding behaviours. In a few exceptions of this species, there are at most rudimentary actions, which are limited to the incubation of the eggs and

Table 3.1 General (interoperative) forms of basic behaviour (MacLean 1975a)

Routinising behaviour
Isopraxic behaviour
Tropistic behaviour
Repetitious behaviour
Reenactment behaviour
Deceptive behaviour

increased attention to the offspring. Crocodiles are the only living reptiles whose behaviour also includes aspects of breeding care (Bellairs 1970).

The special behavioural forms described last are to be distinguished from the following "general" ones (Table 3.1). For the latter, MacLean coined the term "interoperative" forms of basic behaviour, as they are not associated with specific actions but can occur in various situations regardless of the context (MacLean 1975a).

Routinising Behaviour

The term routinisation refers to the ability to structure the daily routine through certain recurring behavioural routines and to modify it by introducing new subroutines. Often, external factors act as triggers for previously rigid behaviours to be adapted. For example, a detour to the sleeping quarters taken once may prove to be safer and is henceforth preferred to the shorter return route.

Isopraxic Behaviour

Isopraxy refers to actions that are performed in the same way (MacLean 1975b). The evolutionary significance of isopraxy, i.e. the fact that certain actions within a species always proceed in the same way, lies in the implicit ability to recognise members of one's own species. Isopraxic behaviour thus also forms part of prosematic communication. The phenomenon can be observed in lizards within a group, for example in the form of head nodding, as a reaction of female and young animals to the appearance of a dominant male. Phenomena of "mass isopraxy" are not found in lizards but find their correlate in the herd movements of turtles. Comparable phenomena in humans, such as mass events or migrations, also seem to have their origin in this basic behaviour.

Tropistic Behaviour

In ethology, tropism describes a negative or positive response to a stimulus.

Unlike the simple reflex behaviour of an animal, tropistic actions, which Lorenz counts among the so-called inherited coordinations (fixed action patterns), require a basic readiness to act. Like reflexes, inherited coordinations represent relatively rigid, easily recognisable, species-specific behaviours, and are extremely stable to the environment, i.e. hardly changeable by external influences (Lorenz 1978).

Noble and Bradley were able to demonstrate in lizards, for example, that when the blue belly region of the males was recoloured to the grey (female) form, courtship behaviour was induced in other male conspecifics (Noble and Bradley 1933).

Repetitious Behaviour

Repetitious behaviour refers to the repeated execution of certain activities. An example here can be the repeated occurrence of recognition and attack actions of lizards. Under certain circumstances, the duration of the repetitions decides more about the outcome of a meeting between two males than the actual physical superiority. As an important part of social communication, repetitions also increase the redundancy of a signal or a message.

Reenactment Behaviour

Compared to repetitious behaviour, reenactment behaviour also consists of repetitive actions, but these are in a meaningful context. Actions that are repeated in the form of a rigid ritual and incorporated into a routine daily structure (daily master routine) can serve as an example here. In the habitual repeated gatherings of animals for mating, repetitive behaviour also serves a survival purpose.

Deceptive Behaviour

Deceptive behaviour reveals an evolutionary strategy for the survival of the individual like hardly any other behaviour. However, this is not only used in dangers but also in food intake or reproduction. The possibilities of deception range from simple mimicry to the so-called death feigning reflex.

All the behaviours described so far can be observed in lizards and mammals. Behavioural patterns that reptiles are not capable of include parental care and the care of offspring, acoustic communication to maintain contact between the mother animal and offspring, and the ability for playful behaviour.

The demonstration of an association between the striatal complex and the behaviours discussed in this chapter requires the experimental elimination of the specific brain region and the simultaneous observation of the resulting behaviour of the animals. One of the first studies on pigeons on this topic was able to show that extensive brain lesions, while maintaining the integrity of the paleostriatum, did not impair the execution of courtship and copulation behaviour (Beach 1951). Subsequent lesion studies on the R-complex of lizards showed that this structure is also essential for the execution of other basic behavioural patterns such as recognition actions and attack actions (Greenberg et al. 1979).

The R-complex and the Behaviour of Monkeys

Ethological studies by Ploog identified four characteristic behavioural patterns in rhesus monkeys: (1) Aggressive behaviour, (2) Courtship behaviour, (3) Greeting actions and (4) Humility or submission actions (Ploog and MacLean 1963).

The aggressive behaviour of the rhesus monkeys shows certain homologies with the territorial or attack displays of the lizards through the threatening gestures. In the attack itself, the monkey screams, approaches the opponent frontally, spreads a leg and rubs his erect penis on the head or side of the opponent (Fig. 3.8).

While the courtship behaviour of the lizards also includes parts of the species-specific attack behaviour, attack and sexual actions in rhesus monkeys are almost identical. The behaviour of a monkey defeated in attack consists of a bending posture of the head or ducking movements of the head and upper body, which corresponds to acts of submission. The already described recognition actions of the lizards are to be seen as a correlate to the greeting displays of the rhesus monkeys. These greeting actions have a characteristic appearance. As soon as an unknown monkey enters an enclosure, the group reacts with immediate climbing up the cage wall, turning the head to the side and retraction of the corners of the mouth. These actions are followed by shrill vocalisations, spreading of the lower extremities and thrusting movements of the erecting penis (MacLean 1964). The fact that rhesus monkeys produce behavioural patterns of aggressive and sexual nature even in front of a mirror image allows for a more precise experimental examination of the effects of a lesion of the

Fig. 3.8 Attack behaviour of a dominant rhesus monkey (Ploog and MacLean 1963). (Copyright: Springer)

R-complex on specific basic behaviours. In a systematic study on rhesus monkeys, MacLean confirmed that the structure of the R-complex is also responsible for prosematic communicative behavioural patterns in mammals (MacLean 1978).

The neuroethological results in lizards and rhesus monkeys underline the importance of the R-complex for the control of prosematic innate actions and the associated psychological aspects.

The Reptilian Brain of Humans

The striatal or R-complex of the reptilian brain represents a large part of the so-called basal ganglia in the human brain and in the mammalian brain (Table 3.2).

Neurological diseases that lead to functional disorders of these structures (such as Parkinson's disease and Huntington's chorea) offer a unique opportunity to gain insights into behaviour-related aspects of the reptilian brain in humans.

Table 3.2 Structures of the basal ganglia in mammals

Basal Ganglia Structures		*Associated structures*
STRIATUM	*GLOBUS PALLIDUS*	
Caudate Nucleus Putamen Ventral Striatum Nucleus Accumbens Olfactory Striatum	External and Internal Segment Ventral Pallidum	Substantia Nigra Ventral Tegmental Area Subthalamic Nucleus

The pathophysiological substrate of Parkinson's disease consists in the degeneration of dopaminergic neurons located in the midbrain substantia nigra with consequent undersupply of striatal structures with dopamine, which leads to the known extrapyramidal-motor disorders in the affected patients. Apart from a few case reports, behaviour-related characteristics of Parkinson's patients have so far been neglected in the literature. At the beginning of the twentieth century, Vogt already pointed out that the striatal complex is involved in the control of expressive movements, e.g. in gestures and facial expressions (Vogt and Vogt 1920). Other authors also confirmed the importance of the basal ganglia for the implementation and communication of internally perceived emotions and needs. The characteristics of Parkinson's patients, which can be described as disorders of the integration of viscerosomatic functions, include dysfunctions of basal behaviours such as eating, defecation or copulation (Yakovlev 1948).

As already demonstrated using the example of lizards, the striatal complex is associated with the execution of basic behaviours in the form of a routine daily structure (daily master routines). Recent studies on Parkinson's patients confirm that they suffer from a disorder of the structuring of the regular daily routine, which is not only due to a pure motor origin (Camara Magalhaes et al. 2005). In addition, these patients also seem to have a specific deficit related to the acquisition of new behavioural habits (Taylor and Saint-Cyr 1995). This ability to acquire new habits is reminiscent of the routinising behaviour already described in lizards, which is also tied to the integrity of the striatal complex. Patients with Parkinson's disease therefore not only show an inability to maintain existing behavioural patterns with the usual daily structure but also to incorporate new subroutines (habits) into the daily structure (Witt et al. 2002). Huntington's chorea is another entity that represents a disease of the extrapyramidal-motor system. In this disease, there is a genetically determined degeneration of the striatal complex. Clinically, in addition to movement disorders, patients primarily show initial neuropsychiatric symptoms. Neuropsychological studies mainly point to an inability of the patients to plan and organise their daily activities. The difficulty lies in the independent initiation and consistent execution of actions, which, however, can be mastered by the patients without problems when externally specified (Caine et al. 1978). Although critics could argue that this could also be a cortical dysfunction, the symptoms would be very consistent with the findings of comparative neuroethology. Here too, malfunctions in the (R-complex controlled) sequence of routine actions and subroutines (habits) could be used as an explanatory model, which would be explainable with the death of the striatal neurons. The scientific investigation of habitual actions and other basic behavioural patterns in Huntington's patients and patients with Parkinson's disease has so far been largely neglected by the neuroscience side, but this would be particularly important from a comparative neuroethological perspective and provide insights into the function of the archaic "reptilian parts" of the human brain. However, recent research results already clearly highlight the importance of the basal ganglia for basic behavioural patterns such as routine actions, repetitive behaviour, habits or ritual actions, thus confirming previous neuroethological data (Saka et al. 2004; Yin and Knowlton 2006; Graybiel 2008).

From the perspective of evolutionary neuroethology, the R-complex or reptilian brain is not an isolated cerebral structure but is highly networked with the limbic system and the neocortex. Accordingly, information about the functioning of the reptilian brain in humans can be obtained not only through lesion studies but also through direct observation of human behavioural patterns. Basic archaic behaviours are therefore woven into rational human behaviour and can be differentiated from these by the method of comparative behavioural research. The cultural process of humans does influence the appearance of these basic behavioural patterns, but it has no influence on the continuing instinctive character of this behaviour. The argument that animal behaviour is fundamentally not comparable to human behaviour can be countered insofar as brain formations that have the same structures, neural connections and organisations in several species are implicitly associated with the same functions. According to the subdivision of basic behaviours in animals, those of humans can also be differentiated into special and general forms.

Special Forms of Basic Human Behaviour

Territoriality

From the perspective of comparative behavioural research, humans as a "territorial species" tend to occupy and claim certain spatial areas (Eibl-Eibesfeldt 1984). This behaviour seems to have gained importance in the course of phylogenesis, especially with the process of human sedentariness, i.e. the transition from a hunting to an agricultural culture. The tendency of humans to establish their own "territory"—ranging from the fencing of the living area, through city walls, to national borders—and to demarcate it from others, cannot be denied. The fact that wars are usually triggered by violations of territories or borders underlines the importance of territoriality as a basic behaviour, even in modern civilisation. But also in individual behaviour, habits such as a person's well-known "favourite place" or the (often not conflict-free) reservation of seats in public areas testify to this basic tendency (Eibl-Eibesfeldt 1999).

The possibilities of "marking" a territory range in the animal world, from setting olfactory scent marks with urine, to visual territorial expressions in rhesus monkeys in the form of rhythmic movements of the erect genitalia. In humans, visual symbols are mainly used to represent territorial boundaries. The habit of many primitive cultures to use so-called house guards in the form of phallus-like stone sculptures to demarcate house boundaries, reminds from a comparative ethological perspective of the previously cited expression behaviour of rhesus monkeys. Even in antiquity, stone phallus sculptures associated with the god Priapus were used for border marking (von Wickler 1966). This observation suggests that this might be an atavistic behaviour pattern, which despite symbolic reworking contains the remnants of animal urogenital territorial actions. In this context, the use of flags—which from a psychoanalytic perspective correspond to an abstract phallus symbol—as a territorial sign of a country also reminds us of the phylogenetic heritage of this behaviour. The most sublime form of human territorial action could probably be the habit of signing a guest book seen as a symbolic gesture of a visual marking.

Threatening or Aggressive Behaviour

A common characteristic of threatening gestures or aggressive behaviour, found in reptiles, mammals and humans, is the enlargement of the appearance. As already described, in lizards, shortly before the actual attack there is a positioning of the two opponents side by side, to demonstrate the full size of the body to the opponent.

Fig. 3.9 Threatening gesture of a chimpanzee. (Image with kind permission from Prof. Irenäus Eibl-Eibesfeldt)

The size of the body is demonstrated in mammals through similar behaviour patterns in the form of an upright body and bristling of the fur. In many mammals, and especially in primates, this behaviour is further enhanced by a threatening facial expression (Fig. 3.9).

In humans, threatening postures show similar tendencies, to enlarge or intensify the appearance, for example in the form of sideways bending of the arms, threatening facial expression or the only rudimentary existing piloerection, which corresponds to the animal bristling of the fur.

The previously described behaviour of genital presentation (Fig. 3.10) in primates is not only part of the territorial expression behaviour but also directly used as a threatening gesture (von Wickler 1966).

In humans, actions of genital presentation were observed in Melanesian tribes. This behaviour is also considered an expression of aggression and dominance and is performed by these tribes mainly in states of threat or fearful excitement in the form of a common ritual dance (Gajdusek 1970). In Papua New Guinea, some tribes use artificial aids, which are put on the penis to enhance the effect of genital presentation. The already described

Fig. 3.10 Genital presentation in primates. (Image with kind permission from Prof. Irenäus Eibl-Eibesfeldt)

attack behaviour in Rhesus monkeys in the form of spreading of the legs and presentation of the erect genitalia is almost identical as a threatening gesture in the Bushmen of the Kalahari (Eibl-Eibesfeldt 1971).

Another example of special basic behaviours is the submission or surrendering action, which opposes aggressive behaviour. A phenomenological feature of these actions, which can be traced from reptiles, through primates, to humans, is the lowering or bending of the upper body or the associated inclination of the head (Fig. 3.11).

The human correlate of submission gestures is rudimentary in the head inclination or bowing during a greeting, or in an evident form such as a ritual submission gesture during the ordination of a priest.

General Forms of Basic Behaviours in Humans

Routinising Behaviour

As already illustrated by the behaviour of reptiles, the term routine refers to the ability to structure the course of the day through typical recurring behavioural routines and to modify it by introducing new subroutines. The tendency to habitually design the daily routine in a stereotypical way is undoubtedly true for many people, as are the associated emotional difficulties, which arise when such routines are broken—for example on holiday or upon retirement. Likewise, many people are familiar with habits that have proven successful in various respects and are subsequently incorporated into everyday actions as a subroutine. Learning processes are naturally involved in the introduction of such subroutines. In a societal

Fig. 3.11 Submission gesture of a chimpanzee. (Image with kind permission from Prof. Irenäus Eibl-Eibesfeldt)

context, forms of routine behaviour can be found, for example, in certain ritual actions, which are also characterised by a predetermined sequence of actions. Since these, for example in the context of ceremonies, are always carried out in a meaningful context, we speak here of repetitive behaviour. Repetitive behaviour in the animal kingdom corresponds to the recurring (seasonal) gatherings of certain animal species for the purpose of mating.

Isopraxic Behaviour

Isopraxy, as already mentioned, refers to imitative behaviour. Natural isopraxic behaviour in humans can be observed in any conversation situation, insofar as there is an unconscious tendency to adopt the body position or posture of the conversation partner. Isopraxies are most impressively demonstrated in certain neurological diseases. The Tourette syndrome, for example, is characterised not only by simple motor tics but also by complex tics in the form of imitations of the movements of others (echopraxia) and the repetition of foreign sounds (echolalia). Although the aetiology of this disease is unclear, a disturbance in the functional loop between the striatum and the prefrontal cortex is suspected. However, imaging procedures confirm previous neuroethological findings that the basal ganglia are also associated with basal isopraxic behaviours (Singer et al. 1993). Against the argument that all imitative behaviour is learned and does not represent an innate behaviour, it can be objected that even highly mentally retarded children are capable of isopraxic actions.

Among primitive peoples, isopraxies seem to be part of the natural behavioural repertoire and also have protective functions. According to ethological reports, when primitive peoples first came into contact with ethologists, all members of the tribe imitated certain actions and gestures of the researchers as if they were thereby performing a symbolic union with the strangers (Gajdusek 1970).

Tropistic Behaviour

Behavioural research defines tropism in the animal kingdom as a negative or positive response to a stimulus, with a tropistic action, in contrast to simple reflex behaviour, presupposing a basic readiness to act. The inherited coordinations (fixed action patterns) that are part of tropistic behaviour are naturally most evident in the human infant, who due to his helplessness relies on instinctive behaviour patterns (Eibl-Eibesfeldt 1999). A typical example of a tropistic behaviour is the smile of the infant, which can initially even be induced by strongly reduced symbolic representations of a face (Spitz 1965). The biological purpose of this behaviour is to appease or placate the counterpart. Observing the behaviour of adults, it is reasonable to assume that a large part of human consumer behaviour is shaped by tropistic behaviour patterns. As motivation research proves, it seems that the advertising industry in particular is geared towards generating motivations or needs that are satisfied by tropistic behaviour. Interestingly, these economic sectors also use brand-specific symbols or "logos", which strongly reminds of a stimulus-response scheme of instinctive behaviour patterns. In addition, the phenomenon of imitation inherent in fashion behaviour or fashion trends seems to correlate with isopraxic tendencies.

Repetitious Behaviour

Repetitious behaviour, in the sense of repeated execution of activities, is found in humans especially in actions that in the animal kingdom are referred to as so-called displacement activities. Tinbergen was able to show that animals in conflict situations, such as during a fight, perform such behaviours in the form of grooming movements that have nothing to do with the original (fighting) mood (Tinbergen 1940). According to the views of some ethologists, pent-up excitement "jumps" from one cerebral centre to another and finds its discharge there. Displacement activities of this kind also manifest themselves in humans in conflict situations, such as during public lectures or performances, which make the person appear "nervous". Typically, this also leads to movements that, as head scratching, beard stroking, mouth wiping, nose rubbing, hair pushing back, etc., remind of grooming movements (grooming behaviour) or body care movements of animals (Seiss 1965). Displacement activities in animals can be triggered by stimulation of the hippocampus with subsequent propagation in parts of the R-complex (MacLean 1957) and show in humans also in the context of temporal lobe seizures, whose discharge patterns can also propagate from limbic areas into parts of the striatum (Wieser 1983).

The R-complex-associated behaviours in humans described so far should not be led to the wrong view that the striatum is an isolated and evolutionarily unchanged structure that "preserves" archaic behavioural patterns in encapsulated form. Rather, comparative-anatomical data show that in the course of the evolution of the human brain, there was a significant increase in the size of the striatum parallel to the expansion of the neocortex. However, the development of the neocortex did not lead to a complete functional overlay of hierarchically subordinate structures.

While the reptilian brain of various species is responsible for basic behaviours and the associated prosematic communication, the evolution of the limbic system was necessary for the ability to perceive the subjective self—a central point in evolutionary neuroethology.

3.3 The Palaeo-Mammalian Brain (Limbic System)

Evolution and Anatomy of the Limbic System

The so-called palaeo-mammalian brain, which emerged with the evolution of the first mammals, consists of a phylogenetically old cortex and its connections with structures in the brain stem. Paul Broca was the first to recognise this development of the forebrain as a specific formation of the mammalian brain and subsequently referred to it as the "great limbic lobe".

The discovery that this is a specific functional unit is due to Paul MacLean, who coined the term "limbic system" (MacLean 1952). The evolutionary transition from reptiles to mammals was characterised by the emergence of specific behavioural patterns linked to the evolutionary development of the limbic system. These mammal-specific behaviours include parental care and offspring care, acoustic communication to maintain contact between mother and offspring, and the ability for playful behaviour.

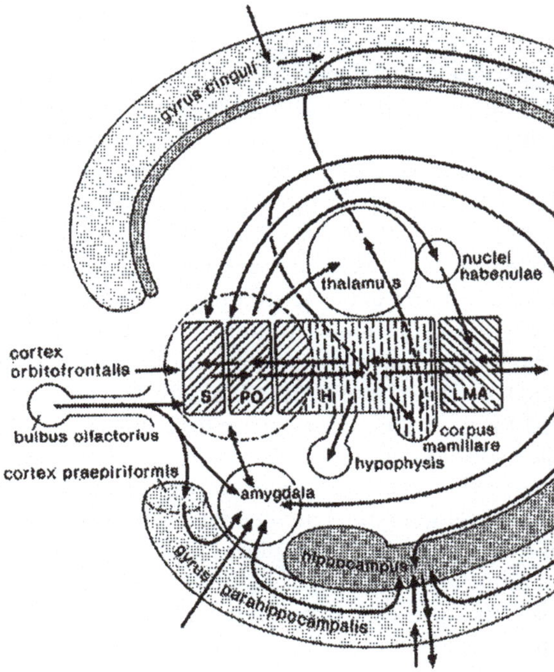

Fig. 3.12 Schematic representation of the limbic system and its connections. H = Hypothalamus, LMA = Limbic midbrain area, PO = Perioptic region, S = Septum. *The Human Central Nervous System* (Nieuwenhuys et al. 2007). (Copyright: Springer)

Anatomically, the limbic system includes cortical and subcortical structures (Fig. 3.12). The cortical structures include the olfactory cortex, the hippocampus, the cingulate gyrus and the subcallosal gyrus. The subcortical components of the limbic system form the amygdala, the septum, the hypothalamus and the anterior thalamic nuclei (Swanson 1987).

The evolution and expansion of the cortex and the limbic system are those essential changes that form the basis for the emergence of the mammalian brain (Striedter 2005). In particular, the neocortex, which is only rudimentary in reptiles, undergoes a special expansion and differentiation in the course of primate evolution. Nevertheless, from a neuroethological perspective, the development of the limbic system appears more significant for the differentiation of mammal-specific behaviours.

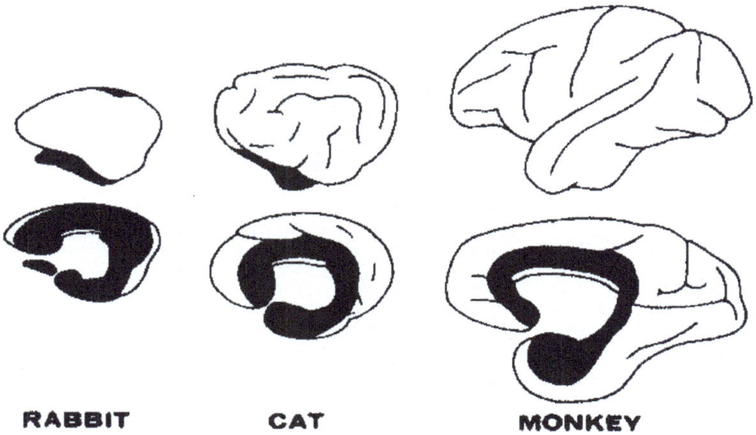

Fig. 3.13 The limbic system of three representative mammals (from left to right: rabbit, cat, monkey). During the evolution of primates, there was a significant expansion of the neocortex compared to the limbic system (MacLean 1954). (Copyright: Springer)

Comparative-anatomical studies show that the limbic system is the common denominator of the mammalian brain (Fig. 3.13).

The lesion and stimulation studies on animals and the clinical data collected on humans imply that the limbic system plays a significant role in the perception and processing of emotions or affects, which are essential for self-preservation behaviours and ultimately for the survival of the species.

Functions of the Limbic System

Understanding the function of the limbic system in mammals for behaviours related to aspects of self-preservation and reproduction requires knowledge of the anatomical peculiarities of this structure. Based on cytoarchitectonic findings, three core groups can be identified within the limbic system that send afferents to the respective limbic cortex areas. These core groups include the amygdala, the septum and the thalamus (Fig. 3.14).

Lesion studies on primates suggest that the amygdala of mammals is essential for self-protection-related behaviours, food choice and the regulation of spontaneous activities (Akert et al. 1961). The removal of the amygdala, including the anterior parts of the temporal cortex, results in

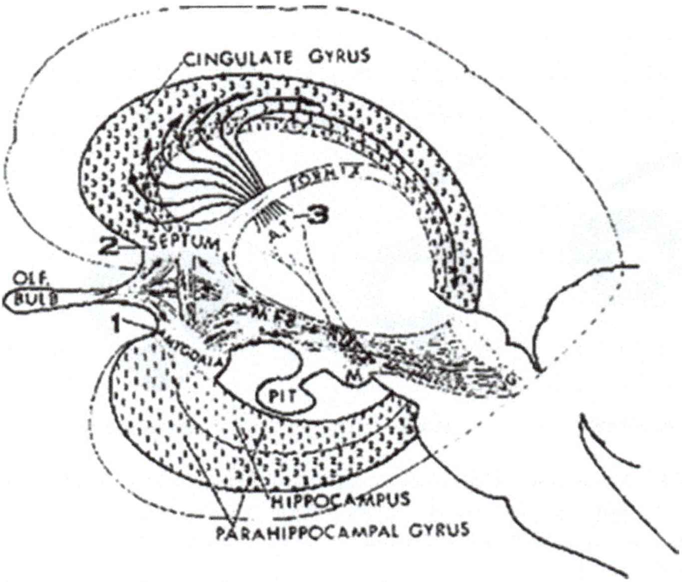

Fig. 3.14 The three parts of the limbic system. (The nuclear groups with the amygdala, the septum and the thalamocingulate system are marked with large numbers (1, 2, 3). The corresponding cortical areas of the limbic system are each represented with small numbers (MacLean 1990). Copyright: Springer)

the test animals no longer being able to differentiate between dangerous and safe situations. This aspect of amygdala function is also confirmed by findings in humans. Patients with bilateral lesions of the amygdala show significant deficits in their ability to recognise negative emotions in the faces of other people—corresponding to an inability to perceive a potentially dangerous situation (Adolphs et al. 1999).

Observations of the social behaviour of primates with such brain lesions revealed that there is also a change in the hierarchical relationship system of these animals, as they lose their dominance behaviour (Plotnik 1968). This would have dramatic effects on behaviour in the wild and underscores the importance of the amygdala for the self-preservation of the individual or the species. The effects of ablation of the amygdala in lower mammals correlate with the previous findings, as these also lead to increased tameness and a "loss of wildness". Among the self-preservation functions that are also mediated via the amygdala are actions that include aggression, fighting and defence, for example, to obtain or defend food.

The importance of the septum, in contrast to the amygdala, lies more in reproduction-relevant aspects of behaviour, such as in the controlling of mating behaviour or the act of copulation. Stimulation attempts on the human septum accordingly induce subjective sensations of well-being, which some patients describe as orgasmic sensations (Heath 1963). The specific orientation of the septum towards reproductive behaviour makes it evolutionarily plausible to assume that this system also provides the biological prerequisite for the emergence of maternal care behaviour in mammals. In both mice and rats, significant disturbances of maternal care behaviour were seen after septal lesions (Slotnick and Nigrosh 1975). Animal experimental investigations furthermore prove that this behavioural pattern is based on a neural network that apparently extends from the septum via the amygdala and the dorsal hippocampus, to the cingulate cortex. Similar functional overlaps are also found for the grooming behaviour of animals, which occurs after actions of food intake, mating, but also after caring or combative activities. Lesion studies showed that for grooming behaviour, both the hippocampus and the septum are responsible (Colbern et al. 1977).

The third functional unit within the limbic system, the so-called thalamo-cingulate system—i.e. the connections of the anterior thalamus and other thalamic structures with the mesocortical gyrus cinguli—seems to play a significant role in elementary sexual functions (Fig. 3.15).

The stimulation of various areas in this region, including the mammillothalamic tract (MT), the medial forebrain bundle (MFB) and the cingulate cortex, induce an erection in monkeys. Stimulation of the spinothalamic tract and the infrathalamic region, on the other hand, lead to genital sensations, up to ejaculation and motor phenomena in the form of genital scratching movements (MacLean 1962). Often these reactions are also accompanied by oral chewing or smacking automatisms, urination or vocalisations. The cingulate cortex also seems to be important for the control of maternal care behaviours, as well as the amygdala and the septum region (Slotnick and Nigrosh 1975; Fleischer and Slotnick 1978; Calamandrei and Keverne 1994). It should not go unmentioned that these actions in lower mammals not only include the rearing and protection of the offspring but also the activity of nest building. The audiovocal communication between the mother animal and the young animals is non-existent in reptiles and is also an achievement of the mammalian brain. A comparison of the so-called separation cry of newborn rhesus monkeys, macaques and human infants yielded the astonishing result that this cry in all cases has the same frequency

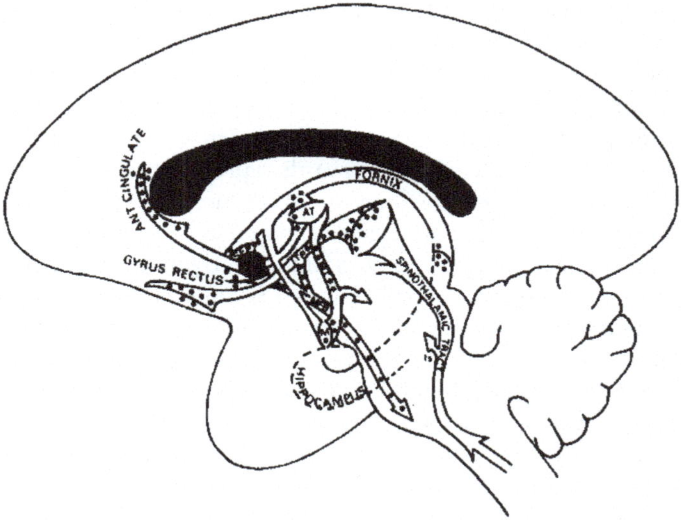

Fig. 3.15 Schematic diagram of the most important structures for the elementary sexual functions in the rhesus monkey. The areas marked with dots or lines represent regions where electrical stimulation can induce erections (MacLean 1962). (Copyright: Springer)

spectrum (Newman 1985). From an evolutionary point of view, this seems to be the first vocal form of communication in mammals, whose neural substrate is also located in the cingulate cortex (MacLean and Newman 1988).

One aspect that has been largely neglected in previous neurobiological research is the specific phenomenon of playful behaviour in mammals. Although playful actions are found in all mammals and in humans, their phylo- and ontogenetic significance is still unclear. There is also currently uncertainty regarding the neural circuits associated with this activity. Murphy et al. described significant disturbances in the play behaviour of hamsters after ablation of the cingulate cortex (Murphy et al. 1981), but several other limbic areas, including the amygdala and hippocampus, also appear to be involved (Panksepp 1998).

The example of play behaviour shows that the behaviour-related functions of the limbic system cannot be reduced to individual structures, but that specific regulatory circuits must be assumed. Current neurobiological research provides a wealth of detailed information about the individual subsystems of the limbic brain, whose fundamental evolutionary role as an archaic system is largely recognised within the framework of the Triune Brain concept (Morgane et al. 2005). With the expansion of the neocortex in humans, the limbic system seems to be increasingly concerned with basal aspects of emotion processing and the transmission of motivation-related information to the phylogenetically younger brain structure, thus ensuring complex human behaviours and a sense of individuality. With this function, the limbic system in humans also contributes significantly to the subjective experience of emotions, the feeling of personal identity and the memory of subjective experiences.

The Limbic System of Humans

Psychomotor Epilepsy and the Limbic System

According to Paul MacLean's "Triune Brain" concept, the human limbic system is crucial in establishing a connection between the "self" and inner experience and external reality. Unlike the basal ganglia, the hippocampus, as a cortical structure (archicortex) in the limbic system, has the ability for increased neuronal synchronisation and excitability, in the sense of an epileptogenic tendency. In particular, processes such as hippocampal sclerosis or atrophy can lead to increased cerebral excitability in this region and thus to epilepsy (Engel 1996). A neural discharge in the context of a seizure event is similar to an electrical stimulation of the brain region that represents the origin of the seizure. In addition, the epileptic excitation can also affect neighbouring limbic structures, in the sense of a so-called propagation. These peculiarities thus provide unique insights into complex human perception and behavioural phenomena that are exclusively associated with so-called partial-complex or psychomotor seizures—and thus with the limbic system. Hardly any other disease is capable of providing such impressive and important information about a brain structure that is concerned with the subjective experience of emotions. This circumstance thus provides valuable insights into paleopsychic processes and aspects of human proto- and emotomentation, referring to the "Triune Brain" concept.

Psychomotor seizures are characterised by two fundamental peculiarities, namely the occurrence of so-called auras and automatisms.

Auras are characteristic conscious perceptions that precede the seizure. One of the most frequently reported aura symptoms by patients is a feeling of nausea rising from the stomach area. Other aura symptoms include feelings of anxiety, psychological phenomena, such as déjà-vu or jamais-vu perceptions, depersonalisations or altered perceptions of the environment in the sense of macropsias or micropsias (Wieser 1983). All these phenomena, which are consciously perceived by the patient, already indicate essential functions of the human limbic system, namely the generation and perception of emotions, the exploration and subjective experience of external reality, as well as the mediation of the feeling of a personal identity or the feeling of self. Upon closer examination of the various qualities of auras, a basic distinction can initially be made between simple sensations and more complex perceptions, each associated with certain affects or emotions. The evolutionary biological significance of affects is certainly to be seen in their function for the preservation or survival of the individual and the species. Based on the fact that affects can only be perceived by an individual as pleasant or unpleasant, but no neutral affects exist, the possibility arises—taking into account the subjectively experienceable affective states of epileptic auras—to categorise human affects. In this context, Paul MacLean proposed a concept of affects based on neuroethological data (Fig. 3.16), in which he, in addition to the division into pleasant and unpleasant emotions, also made a differentiation between basic, specific and general affects (MacLean 1970).

According to these views, basic affects refer to perceptions of internal or physical states, or needs, such as the need for food, water or sexual contact. The specific affects, which can be innate or acquired, are

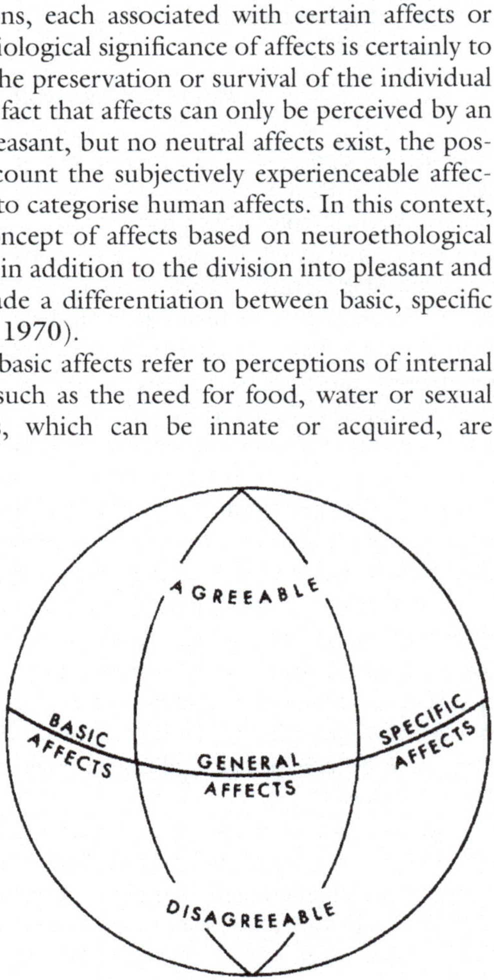

Fig. 3.16 A scheme for the systematisation of affects (MacLean 1970). (Copyright: Springer)

associated with perceptions of the outside world and the associated sensory perceptions. The innate specific affects would therefore include emotional states that are associated with pleasant or unpleasant smells or noises, and whose evolutionary significance for survival is evident. The acquired specific affects of humans could include feelings associated with the perception of music or art.

The so-called general affects, on the other hand, are not tied to an immediate perception and can be effective retrospectively due to cognitive processes, or can recur independently of external occasions. Characteristically, general affects can be triggered by other individuals, situations or just by cognitive processes, such as memories. In addition, there are no specific sensory triggers that give rise to these affects. Phenomenologically, general affects can thus be assigned to emotional states that are referred to as "actual" emotions.

Psychomotor Seizures and Basic Affects

Alimentary Phenomena

The previously mentioned epigastric aura is one of the most common sensations described by patients during a psychomotor seizure. The epigastric perception itself is described as an unpleasant physical stimulus, the accompanying affect almost always has a character of fear. Furthermore, accompanying nausea is reported, which is characterised as rising from the stomach area. Analogous to these phenomena of psychomotor seizures, which can be attributed to basic affects, patients may also experience feelings of hunger or thirst during the aura (Cascino and Sutula 1989).

More recent studies on patients with such seizures even suggest that peri-ictal water drinking or the feeling of thirst is localised in the non-dominant temporal lobe (Trinka et al. 2003).

Cardio-Respiratory Phenomena

Manifestations of the cardio-respiratory system are among the visceral sensations in the context of seizures, which also originate in the medial temporal lobe. The cardiac symptoms manifest themselves mostly in the form of feelings of "racing heart" or palpitations and tachycardias, which occur as an aura without external triggers. Breathing-related ictal phenomena include feelings of shortness of breath to the feeling of suffocation. Naturally, auras with cardio-respiratory symptoms are also associated with feelings of fear.

Urogenital Phenomena

Ictal urinary urgency is another, albeit rare, aura sensation, which can be associated with basic affects. Even with this visceral manifestation, the origin of the seizure appears to be localised in the non-dominant temporal lobe region (Baumgartner et al. 2000; Loddenkemper et al. 2003). The importance of the limbic system for the perception of sexual arousal, up to orgasm, is reflected in the phenomenon of the orgasmic aura (Jansky et al. 2004; Aull-Watschinger et al. 2008).

Psychomotor Seizures and Specific Affects

Specific affects correspond to perceptions of the outside world in the form of sensory perceptions. Auras can be associated with a variety of specific sensory phenomena. Olfactory and gustatory auras, for example, induce smell and taste sensations, which are usually perceived as unpleasant. Acoustic and vestibular symptoms are often associated with more complex emotional states. Thus, a patient perceives a sound or noise not only as such but also associates it with a threatening character. Ictal voices can under certain circumstances hardly be distinguished from actual acoustic hallucinations. Similarly, ictal vestibular sensations can take the simple form of a feeling of dizziness, or lead to complex spatial and scenic perceptions. Visual auras often occur in the form of micropsia or macropsia of the surroundings, less often there can be complex visual hallucinations. In addition to these specific sensory perceptions, there are also reports about ictal feelings of change, which affect the entire body, or about phenomena that can be considered as an isolated symptom of the autonomic nervous system (e.g. feelings of warmth or heat or isolated piloerection).

The presentations mentioned here clearly show that both the enteroceptive and the exteroceptive sensations are associated with corresponding affects, insofar as enteroceptive perceptions are associated with basic affects and exteroceptive perceptions are associated with specific affects. However, no sensation leads to an "indifferent" affect.

Psychomotor Seizures and General Affects

General affects differ from specific and basic affects in that they can refer to situations, individuals or things. They are also characterised by the fact that they can be induced by both physical perceptions and external stimuli. General affects are also independent of current stimulation and can therefore persist or recur, for example by being reactivated through cognitive processes. MacLean points out that certain forms of behaviour are also associated with specific general affects (MacLean 1970). Accordingly, observable search- or appetitive behaviour would be associated with a feeling of desire or longing. Such emotional states are often reported by patients with psychomotor seizures as part of an aura, such as the feeling "of wanting something specific" (Williams 1956).

Another behaviour, aggression, would be associated with the subjective feeling of anger. Further, protective actions can be cited, which correspond to the feeling of fear; separation behaviour, which includes feelings of abandonment and grief; as well as triumphant and caring behaviour, which are associated with feelings of happiness or affection. For almost every emotion just described, there are corresponding affective correlates or sensations in the context of psychomotor seizures.

In this context, so-called indeterminate affects should not be left unmentioned, which due to their characteristics cannot be assigned to the basic, specific, nor the general affective states. Indeterminate affects include feelings of familiarity or strangeness—which may represent two poles of a category—as well as altered perceptions of time and space. There are also corresponding aura sensations for these emotional perceptions.

The phenomenology of psychomotor auras with the corresponding isolated affective perceptions—without any causal connection with cognitive processes—provides impressive evidence for the dissociation between cognition and emotion, and thus between the Palaeo- and Neo-Mammalian brain, already present at the neuronal level.

Psychomotor Automatisms

The aura is usually followed by a disturbance of consciousness, which can also be the only seizure manifestation. However, in many cases, motor phenomena also occur in the further course of the seizure. These so-called automatisms or motor templates manifest themselves, for example, purely somatomotorically in the form of oro-alimentary automatisms as chewing, smacking or swallowing movements. The motor phenomena can also appear in the extremities and lead to fumbling movements, wiping movements or gesticulations as hand automatisms (Engel et al. 1997). Somatovisceral automatisms include cardiovascular symptoms (tachycardia), respiratory and vocal manifestations (coughing, hyperventilation, screaming, whistling, etc.), or thermoregulatory phenomena (goosebumps, trembling, sweating). In a broader sense, urogenital automatisms can also be mentioned in this context. In addition to automatisms in the form of urination, with the adoption of the corresponding body postures, there are also descriptions of genital automatisms during psychomotor seizures, ranging from genital fidgeting to masturbatory actions and thrusting pelvic movements (Leutmezer et al. 1999; Dobesberger et al. 2004).

More complex automatisms can occur in connection with the auras already described (especially with the associated general affects), such as search movements, aggressive or tender actions. Complex automatisms can sometimes even take the form of partly coherent sequences of actions, as in patients who continue to steer vehicles or carry out simple tasks during a seizure. However, they do not remember these activities postictally.

In particular, the observations of complex ictal automatisms raise the question of whether these manifestations are innate or acquired behavioural patterns. Ictal behavioural automatisms, such as the not infrequently observed drumming of fists against the chest, strikingly remind us of the behaviour of primates and suggest an innate atavistic gesture. But also automatisms, such as throwing one's own body against the wall, could be interpreted in this respect as archaic innate behavioural patterns as phylogenetic heritage of human ancestors once serving as a protective reflex (e.g. as a liberation action after falling into a pit). Automatisms in the form of undressing also appear from this perspective as a possible innate primitive reflex to free oneself from "constrictions" or captivity (MacLean 1990).

The hierarchical structure of Paul MacLean's Triune Brain concept provides new and unexpected insights into previously unexplained clinical phenomena. With MacLean's concept and the known manifestations of psychomotor epilepsy, certain human actions can be interpreted and understood as suddenly occurring atavistic or regressive phenomena, the origin of which is to be found in the evolutionarily old limbic system. One such clinical phenomenon is the "Limbic Psychotic Trigger Reaction" (LPTR) described by Pontius, which can be seen as a subtype of psychomotor seizures (Pontius 1993). The LPTR is characterised by suddenly occurring actions associated with autonomic symptoms and a transient psychosis. The actions usually begin and end suddenly, are emotionless, and consistently show seizure-specific sequences in the form of aura, seizure and postictal phase. The actions carried out during such "behavioural seizures"—which in the cases documented so far mostly consisted of murder—are typically directed against strangers who represent a specific stimulus for the perpetrator. The individual stimulus seems to trigger memories of previously experienced stress situations in the perpetrator. Particularly typical are the lack of motive for the act, the ego-dystonic character, as well as the preserved consciousness and the lack of amnesia for the action. This is possibly explained by the lack of involvement of temporal neocortical areas, in contrast to ictal propagation into these structures in the context of typical psychomotor seizures. The LPTR also seems to be associated with a transient abolition of fronto-limbic connections, which would explain the psychosis-like symptomatology of the actions. From a neuroethological perspective, such complex and organised actions could also be considered as atavistic or archetypal behavioural patterns, which belonged to the normal behavioural repertoire in ancient times and were inhibited over the course of phylogenesis through increasing fronto-limbic interactions (Pontius 2002).

Electrical Stimulation and Lesions of the Limbic System

The electrical stimulation of certain brain regions provides insights into the functioning of the corresponding areas. At the cortex, these induced phenomena range from motor to sensitive and sensory phenomena. The first description of mental phenomena induced by stimulation of the human temporal lobe was achieved by the Canadian neurosurgeon Wilder Penfield (Mullan and Penfield 1959), who was also responsible for the experimental proof of the somatotopy of human sensorimotor cortex areas. He confirmed the assignment of body regions to certain cortex areas postulated by Hughlings Jackson. The first phenomena evoked by stimulation of the temporal lobe included complex hallucinations, déjà-vu experiences, feelings of fear or amnesia. The later specific stimulations of the hippocampus and the amygdala in humans yielded very similar results. The majority of patients also reported about induced hallucinatory images or scenes, about emotions of fear or dread, visceral or epigastric sensations, as well as about diverse elementary sensory perceptions of visual, auditory or proprioceptive nature (Halgren et al. 1978). These phenomena correlate exactly with the described perceptions of patients during psychomotor seizures, whose seizure origin also lies in this area of the limbic system. About the mechanism, how a punctual electrical stimulation in this brain area not only leads to individual sensory perceptions but also to complex mental perceptions (hallucinations, dreamy states, déjà-vu, jamais-vu), can still only be speculated about. Several authors hold the view that the induction of complex mental and emotional contents by stimulation at a specific area suggests that this region has access to a

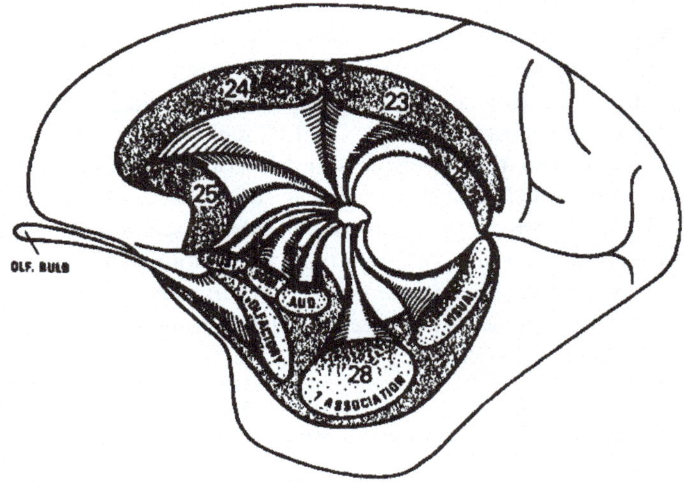

Fig. 3.17 Medial section through the brain of a rhesus monkey with representation of the individual sensory afferents (biocones) to the limbic system (MacLean 1990). (Copyright: Springer)

diffuse, widely branched neural representation in the cortex. The phenomena associated with psychomotor seizures also suggest that the limbic system has connections to all intero- and exteroceptive systems. The phenomena of psychomotor seizures and electrical stimulation attempts on the temporal lobe imply that the cortical sensory pathways to the limbic system have a similar topology as those sensory afferents that project from individual thalamic nuclei to specific cortex areas. The pathways emanating from circumscribed sensory cortex areas would therefore be imagined as cone-shaped afferents ("biocones") that connect with the respective limbic structures (Fig. 3.17). Since the hippocampus receives afferents from almost all limbic areas, it forms an integration centre for both intero- and exteroceptive information (MacLean 1975c).

The integration of information from intero- and exteroceptive systems and the associated influence on vegetative, visceral and emotional functions assign the hippocampus a special role in the development of a sense of individuality and the associated aspects of memory. The so-called Theta oscillations of the hippocampus appear, according to recent studies, to be the neurophysiological substrate of memory processes or memory consolidation (Buzsaki 2002).

Lesion studies of the human limbic system confirm its importance for the integration of signals originating from intero- and exteroceptive systems. Hebben and colleagues (Hebben et al. 1985) reported on a patient who, after resection of both temporal lobes due to therapy-resistant epilepsy, developed an inability to perceive internal states or needs, or to report on them. Compared to other patients with global amnesia, Hebben's patient showed significant limitations in his ability to perceive and interpret physical needs and perceptions such as hunger, thirst or pain, as if he had less access to this interoceptive information. It could thus be shown that this deficit is not due to a malfunction of the memory but is associated with the absence of the amygdala.

Patients with a rare genetic syndrome, Urbach-Wiethe disease, offer the unique opportunity to study the function of the amygdala (Urbach and Wiethe 1929). In a small percentage of patients, there is a selective calcification of the amygdala on both sides. In experimental studies, these patients showed a significant deficit in their ability to recognise emotional expressions in the faces of other people, or to interpret them affectively (Adolphs et al. 1994, 1999). This inability correlates with the previously postulated function for the limbic system insofar as it is an exteroceptive disturbance of the perception or processing of emotional signals. Another report on a patient with bilateral selective amygdala lesion demonstrates that the amygdala, with its connections to the hippocampus, plays a crucial role in storing and remembering autobiographical contents. In the neuropsychological testing of the patient, there was a striking finding in that he had preserved memory functions for autobiographical facts, but an isolated disturbance for remembering autobiographical episodes (Wiest et al. 2006). Autobiographical episodes differ from facts in that the former are associated with emotional contents. A number of further studies support the assumption that the amygdala is essential for the consolidation of emotional autobiographical contents (McGaugh 2004; Buchanan et al. 2005). It seems as if this ability to store and remember emotionally linked experiences (of "ongoing experiences") contributes significantly to the sense of individuality in humans (MacLean 1969).

3.4 The Neo-mammalian Brain (Neocortex)

The Evolution of the Neocortex

The synonym for the cerebral cortex, "Pallium" (cloak), graphically emphasises the anatomical fact that the cortex envelops the subcortical brain structures like a cloak. Comparative-anatomical studies show that all vertebrates possess a—often only rudimentary—cerebral cortex (Fig. 3.18).

Fish and reptiles, in this regard, have a cortex plate consisting of only three layers, which mainly receives afferents from the rhinencephalon and has other sensory and somatosensory connections. It is only in mammals that there is a significant expansion of the cerebral cortex; at the same time, the thalamus and the thalamo-cortical connections significantly increase in size and thus become the most important afferent system for the cortex (Jerison 1973). Taking these anatomical circumstances into account, the term coined by MacLean for the neo-mammalian brain includes both the neocortex and the thalamic connections associated with it (MacLean 1990). In contrast to the limbic system, with its cortical components, there was a massive expansion of the neocortex during evolution, which ultimately culminated in the development of the human cerebral cortex. Due to its extensive networking with a multitude of sensory systems, such as the visual, somatosensory or motor system, as well as the acoustic or the vestibular system, the neocortex is primarily concerned with exteroceptive tasks. The unique ability for linguistic communication also allows for the first time in the history of evolution the interindividual transmission of thought content and the associated subjective sensations or internal states. This new achievement of the cerebral cortex laid the foundation for abstract information transmission between individuals and paved the way—ultimately over generations—for socialisation and the creation of culture.

Since the previous explanations mainly dealt with the concepts of evolutionary neuroethology, here, in addition to cognitive aspects, especially those neocortical functions should be considered that are related to the previously discussed concepts of paleopsychic processes and prosematic communication. Paleopsychic processes include the protomentation and the emotomentation. The term protomentation refers to psychic functions that are involved in the regulation of basic behavioural routines and certain behavioural patterns of non-verbal communication. The concept

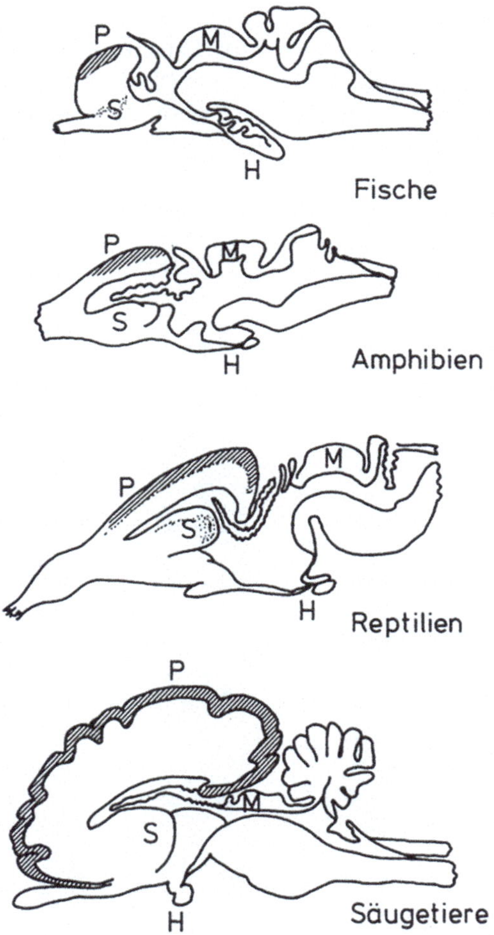

Fig. 3.18 The phylogenetic development of the cerebral cortex using the example of various vertebrates. (A definitive neocortex (hatched) is only found in mammals. Fische = fish, Amphibien = amphibiens, Reptilien = reptiles, Säugetiere = mammals, H = Hypophysis, S = Striatum, M = Midbrain, P = Pallium (Creutzfeld 1983). Copyright: Springer)

of emotomentation, on the other hand, describes a psychic activity that regulates or influences behaviour based on the subjective perception of emotions. In analogy to these non-verbal functions, the "ratiomentation" would thus correspond to the rational, self-conscious activity of the human neocortex. The specific functions and characteristics of the R-complex ("protomentation") and the limbic system ("emotomentation") outlined in the previous chapters raise the question of the significance of the neocortex for these phylogenetically older brain structures.

According to current ideas, it is only the neocortex of the frontal lobe that is responsible for the interaction of cognitive and proto-, as well as emotomental contents (Fig. 3.19).

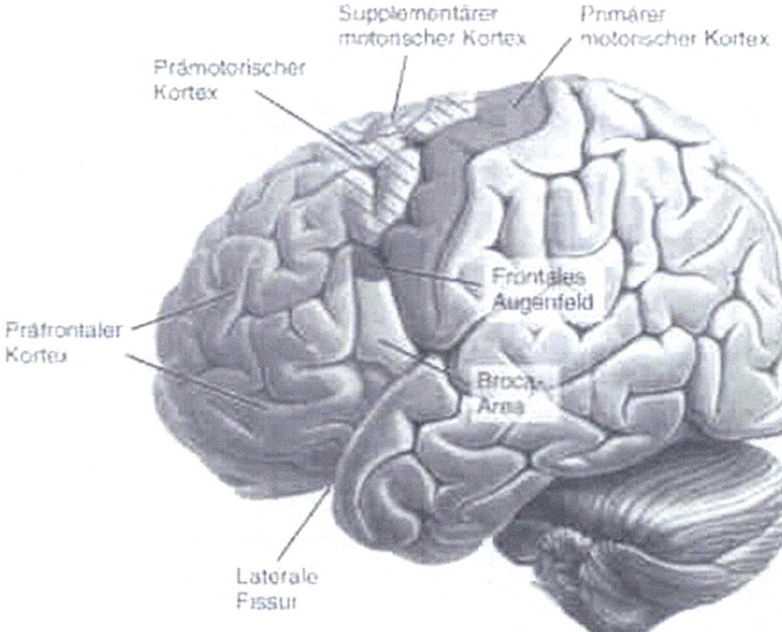

Fig. 3.19 Structure of the functional regions of the frontal lobe in the human brain. (Laterale Fissur = Lateral fissure, Präfrontaler Kortex = Prefrontal cortex, Prämotorischer Kortex = Premotor cortex, Supplementär motorischer Kortex = Supplementary motor cortex, Primär motorischer Kortex = Primary motor cortex, Frontales Augenfeld = Frontal eye field, Broca Area = Broca area (Karnath and Thier 2003). Copyright: Springer)

The evolutionary view of an expanding, especially frontal, neocortex with the corresponding connections to the archaic brain structures of the striatum and the limbic systems provides new insights and explanatory models into previously unsolved phenomena of human experiences. Accordingly, aspects of neural and evolutionary mechanisms of specific human behaviours will be discussed here.

Laughter and Crying

Neurological Aspects of Laughter

Laughter is one of the fundamental behavioural phenomena of humans. In connection with humour and joy, laughter is one of the most inherent human characteristics. In contrast to spontaneous laughter, which is associated with humour or emotional stimuli, the so-called symptomatic or pathological laughter can be observed in patients with certain brain lesions, or with so-called gelastic epilepsy, or induced by electrical cortical stimulation. Such phenomena provide unique insights into the neural mechanisms of laughter and the feeling of joy. One of the main features of pathological laughter is the absence of the correlating affect and the voluntary control over the act of laughter (Poeck 1969).

The brain structures that have so far been associated with pathological laughter originate from the frontal cortex and extend to the ventral parts of the midbrain and the brainstem. The role of the frontal lobe, especially the prefrontal cortex, as a central control area of laughter, is well documented by lesion studies in humans (Mendez et al. 1999; Zeilig et al. 1996). Affectless laughter has also been described in subcortical brain damage, especially in lesions in the area of the striatum and the internal capsule (Ceccaldi and Milandre 1994; Kim and Choi-Kwon 2000), as well as in paramedian brainstem infarctions (Kataoka et al. 1997). The importance of the cerebellum for laughter is still controversial. In analogy to its function in motor skills, some authors have also assigned the cerebellum a coordinating and modulating ability for laughter (Brown 1967; Parvizi et al. 2001).

Electrical stimulation of the cortex is another method to gain information about the neural network of laughter. In this regard, reports from human stimulation experiments show that the stimulation of the frontal cortex and the amygdala induces smiling (Fish et al. 1993). Laughter in connection with a corresponding positive affect could be elicited in two patients by cortical stimulation of the fusiform and the parahippocampal gyrus, with the patients showing a smiling facial expression and reporting a feeling of joy (Arroyo et al. 1993). Emotional laughter could also be induced by electrical stimulation of frontal cortex areas, responsible for the execution of motor programmes, the superior frontal gyrus and the supplementary motor area (SMA, Fried et al. 1998; Krolak-Salmon et al. 2006). With regard to laughter induced by epileptic seizures, the so-called gelastic epilepsy, it should be noted that these seizures usually originate in frontal, temporal or hypothalamic areas.

In contrast to the reflexive character of laughter, humour requires a combination of several higher systems or cognitive functions. To capture a humorous content, in addition to the activation of attention and memory systems, it also requires the ability for emotional evaluation and verbal abstraction. Studies on patients with focal brain lesions confirm that the frontal parts of the non-dominant hemisphere are essential in understanding and perceiving humour (Gardner et al. 1975; Shammi and Stuss 1999). The emergence of new imaging techniques, such as functional MRI, contributed significantly to the identification of further brain areas associated with humour. Currently, a neural network is assumed, which predominantly includes cortical areas (the middle and inferior frontal gyrus, the middle and inferior temporal gyrus, the orbitofrontal and medial prefrontal cortex, the Broca area, the SMA, the visual association cortex and the anterior temporal cortex) but also the putamen and the cerebellum (Goel and Dolan 2001).

In summary of the current neurological knowledge, a picture emerges in which laughter can be considered a human behavioural phenomenon that includes motor, affective and cognitive components. According to the current data situation, especially frontal and temporal structures, as part of a neural network, are involved in the perception of humour. Dorsal brainstem regions are responsible for the activation of the facial muscles associated with laughter. These reactions are inhibited by ventral brainstem regions, which in turn are influenced by frontal areas (Wild et al. 2003). In such a model, cerebellar structures would automatically adapt the act of laughter to the cognitive and situational context.

Neuroevolutionary Aspects of Laughter

In the understanding of evolutionary neuroethology, laughter could have evolved from primitive behaviours of human ancestors (Panksepp 1998). Darwin already described laughter-like vocalisations in chimpanzees and pointed out the similarities with human laughter. Panksepp (2005) was able to demonstrate that even rats emit vocalisations in the ultrasonic range during play or other positive interactions. Since these reactions can also be triggered or strengthened by tickling the animals, the assumption is close at hand that here neural and functional homologies to human laughter exist. The assumption that laughter is an old instinctive and possibly innate behaviour is supported by the fact that children develop the ability to laugh or smile at the age of four months, even if they were born blind and deaf. For the same reason, a distinction must be made between the spontaneous joyful vocalisations of toddlers and animals during social interactions and the humour-associated laughter of adults. From a phylogenetic perspective, laughter could also have served other functions of human communication and socialisation. For some authors, laughter represents not only the positive affective state of an individual, but it can also actively influence the emotional sensitivity or behaviour of a listener (Owren and Bachorowski 2003). According to these ideas, laughter is a species-specific behaviour that signals positive emotions and simultaneously prompts other individuals in a group to associate positive feelings with the speaker. This effect aims to induce the same joyful state in the listener and create a positive context for social interactions.

Phenomenologically, it is of interest that ethology also distinguishes between social smiling and laughter (Stroufe and Waters 1976). In this context, it would be conceivable that smiling (due to the visibility of the teeth) has evolved from archaic threat gestures of mammals. The human smile would thus originate from these old pre-existing forms of interaction and thus communicate that one is generally friendly but always prepared for possible disputes.

According to Panksepp, laughter has evolved from another cerebral network, namely the system responsible for play (PLAY-system, Panksepp 1998). The origins of laughter would thus be traced back to phylogenetically old components of play behaviour.

Another neuroevolutionary concept, regarding the origin of laughter and humour, is based on a theory of "false alarm". According to these views, both laughter and humour are characterised by an increase in expectations. These expectations are subsequently interrupted by a sudden unexpected turn (a funny event or punchline). As long as the unexpected turn proves to be non-threatening, this also leads to a decrease in expectations. The loud and explosive vocalisation serves to inform others that it is a "false alarm" that they do not need to pay attention to (Ramachandran

1998). An example should illustrate the mechanism more precisely: If a person seriously injures themselves in a fall, an observer usually does not react with laughter but rather feels compelled to rush to the person's aid. However, if the person does not injure themselves (this represents the standard situation for every slapstick humour), the observer feels a compulsion to laugh, thereby signalling to other observers that they do not need to rush to help (corresponding to an "all clear"). In this view, laughter appears as a social signal of the "false alarm" (Ramachandran 2003). Consequently, jokes would be attempts to minimise or trivialise signals or content that would normally represent a disruption or serious danger. In this light, one would be tempted to establish a correlation between human cognitive evolution and the development of humour and wit, by a mechanism that originally served as an ethological signal to alert members of a social group, which was then internalised in the course of phylogenesis and is now exclusively concerned with cognitive aspects or anomalies, in the sense of a psychological defence mechanism. The development of more differentiated cognitive functions in humans ultimately also led to increased (not conflict-free) interaction with the instinctive components of behaviour, so that with the evolution of humour, an old behavioural instrument is occupied by internalisation, less with external signals, than with internal conflicts. From a systems theoretical perspective, a former closed biological programme has thus become an open programme.

Neurology and Evolution of Crying

As an expression of a disinhibition phenomenon, crying, like laughing, can be observed in patients with brain lesions (Poeck and Pilleri 1963). Since this so-called pathological crying is often associated with "pathological laughter", or both phenomena can merge into one another in a patient, a common cortical control of these behaviours is to be assumed. The triggering brain damage is usually found in the area of the internal capsule, in the basal ganglia, or in thalamic or hypothalamic regions (Poeck 1985). Clinically, the affected patients respond to unspecific external stimuli with uncontrollable, sudden onset and uncontrolled crying, with corresponding acoustic, respiratory and vegetative phenomena. Pathophysiologically, it is assumed that there is a disinhibition of pontine and mesencephalic control centres, which are subject to the control of the anterior cingulate cortex. In rare cases, pathological crying has been reported in the context of deep brain stimulation. In a patient with Parkinson's disease, pathological crying could be induced by stimulating the subthalamic nucleus of the right hemisphere. Since in this patient, during stimulation in the positron emission tomography (PET), activations were also shown not only subthalamically but also ponto-cerebellar, the latter regions also seem to be involved in the psychomotor control and modulation of crying (Wojtecki et al. 2007). Compared to gelastic epileptic seizures, ictal crying is a rare event. In the seizures documented so far, the ictal activity was found exclusively in the right hemisphere, which is apparently important for the processing of negative affects (Luciano et al. 1993).

In the discussion of evolutionary aspects of crying, the specifically human ability for tear secretion requires special consideration. Although some primates display elements of crying, or correlating vocalisations in their behavioural repertoire, humans are the only creatures capable of crying with tear secretion. Ontogenetically, it is noteworthy that the so-called scream crying of the newborn is initially still tearless. Some authors therefore assume an evolutionary advantage in the ability to secrete tears, as the infant, who has become increasingly dependent in the course of human evolution, received better vision and thus protection from external dangers through the tear fluid (Montagu 1959). MacLean pointed out that the use of fire is reserved only for humans, and here a possible connection to the evolution of tear secretion exists. He sees a connection between the use of fire in the dwellings or caves and the associated smoke development and induced tear secretion, as well as the social gatherings and communications of our human ancestors around the fire (MacLean 1987). These gatherings could be seen as favourable factors for the increasingly differentiated social and emotional relationships of early humans, and it would be plausible that in this context the origin of the exchange of negative emotions and human grief would be found.

Handedness and Language

The transition from the prosematic mode of communication of the limbic system, with its undifferentiated affective vocalisations, to human language, from an evolutionary perspective requires not only an expansion of the frontal neocortex but also an increasing lateralisation of brain functions. Indeed, the human brain exhibits a higher degree of asymmetry than that of primates (Preuss 2004). Anthropological studies suggest a relatively late acquisition of language (about 100,000 years ago) in the phylogeny of humans (Mellars 1998). The evolutionary developments of handedness and human language seem to be closely linked. The genetic aspect of language evolution has only come into focus in cognitive neuroscientific research in recent years. An isolated gene in this regard, FOXP2, which differs between humans and chimpanzees, has been associated with oromotor coordination or articulation (Enard et al. 2002).

From a neuroevolutionary perspective, there are a multitude of explanatory models for the lateralisation of language and handedness. In contrast to language, the prosematic vocalisations originating from the limbic system, which naturally occur primarily during affective outbursts of pain, fright, triumph or pleasure, do not seem to depend on cerebral asymmetry and possibly also not on the function of the neocortex itself. Anatomical conditions also represent an important factor in the lateralisation of language control. Penfield already pointed out in the context of stimulation experiments of the motor cortex that unilateral electrical stimulation of a hemisphere always induces a contralateral movement corresponding to the cortical area, except for organs, such as the tongue or the pharynx, which are arranged in the midline (Penfield and Welch 1951). Indeed, a bilateral motor innervation of an unpaired, midline-near organ seems neurologically redundant, so a unilateral control would be of evolutionary advantage. Given the fact that about 95% of people are right-handed, the question arises as to the reason for the shift of motor dominance to the left hemisphere. A theory that explains the preference of the left hemisphere, both for the language function and for handedness, refers to the activities of early humans as hunters in groups. Since undifferentiated prosematic vocalisations during group hunting would be disruptive, and a coordinated communication among the participants is necessary for such a complex task, it seems plausible to see in this activity a favourable factor for the development of a precise means of communication, the control of which has now been shifted to the neocortex. Since the skilful use of a weapon is of paramount importance for hunting, it would also be evolutionarily advantageous to shift motor and language functions to the same hemisphere to ensure the fastest possible and effective coordination.

Another hypothesis regarding the origin of handedness refers to the proven habit of women to carry their infants on the left arm (Salk 1960). It is assumed that holding the child on the left side is associated with a calming effect on the child as it can better perceive the mother's heart sounds. As a result, over the course of evolution, the right hand increasingly became available for other functions and thus became the hand used for more specific motor activities. The calming effect for the infant, however, not only has individual positive emotional effects but would also have provided survival advantages in phylogenesis, as a screaming child in dangerous situations would have posed a greater risk to the family and community (MacLean 1985).

Anticipatory Behaviour, Altruism and Empathy

Among the higher cognitive achievements of humans, the ability for insightful and anticipatory behaviour is certainly one of the most significant evolutionary achievements. Like few other abilities, this characteristic has enabled humans to shape the environment and ultimately create cultures. The assumption that this behaviour is linked to neocortical frontal functions is supported by clinical findings from patients with bilateral lesions of the prefrontal cortex, as these patients show striking deficits in the planning and memory of action sequences. The actions of the patients are more like those of toddlers, whose immature prefrontal structures do not yet allow for anticipatory action. Action options are usually designed taking into account past experiences and the current context of the situation. In cases where there is no experience regarding a possible problem solution, or if the current situation does not suggest an optimal option, the acting individual relies on their ability to plan ahead. This can also be referred to as anticipation ability or intuition, i.e. the ability to rely on certain expectations. The ability for such behaviour can be experimentally evaluated by examining the solution approaches of individuals in maze tasks. Patients with frontal lesions showed clear disturbances in their anticipatory or intuitive behaviour in complex decision-making situations (Karnath et al. 1991). Similarly, such patients showed a tendency towards incomprehensible solution hypotheses in tasks that require the ability to consider rules and concept formation (Burgess and Shallice 1996).

As already outlined by MacLean in the "Triune-Brain" concept, it seems unlikely that even the highest cognitive achievements of the neocortex—such as forward-looking and anticipatory behaviour—can be carried out entirely without interaction with evolutionarily older brain regions. Damasio et al. postulate that so-called somatic markers are included in all cognitive decision-making processes (Damasio et al. 1991). According to this theory, the somatic markers, which encompass all visceral (also referred to as "gut feeling" in certain decisions) and non-visceral processes in the body, contribute to evaluating all possible alternatives of actions and ultimately choosing a successful, constructive and safe action option for the individual.

Limbic areas, which have extensive connections with the prefrontal cortex, are suspected to be the starting point of these markers. Damasio's assumption that these limbic components come into play particularly when a rapid rational decision is not possible, correlates with the postulate of evolutionary neuroethology that the limbic system is also involved with survival strategies and tasks related to species preservation. Altruistic, or selfless, behaviour is a phenomenon that can be observed in animals. The so-called reciprocal altruism between unrelated animals is characterised by a fundamental reversibility of the relationship between the altruist and the beneficiary, so that the altruist is likely to have the service returned by the beneficiary at a later date. This behaviour is compatible with the principles of natural selection from an evolutionary perspective. The phenomenon of so-called true altruistic behaviour, i.e. selfless actions without expected reciprocation, was only explained by the introduction of the concept of "kin selection" in ethology. This form of altruism benefits related individuals exclusively, and while it does not lead to an immediate survival advantage for the selflessly acting animal, it does contribute to the survival of its own genes through the indirect route of relatives. The ethological perspectives and explanatory models of altruism can also be applied to human selfless behaviour (this is evident in well-known phrases such as "one hand washes the other", or in the principle of "nepotism"). However, in the case of a helper who comes to the aid of a stranger who is injured, ethological or biological approaches are no longer sufficient to explain this behaviour.

Instead, acquired prefrontal abilities come into play, as this course of action is now expected by the cultural environment of the helper. Such human behaviours thus represent the pinnacle of culturally shaped social interactions, which, while not biologically adaptive, are culturally adaptive.

Empathy is a specifically human achievement. The behaviour seems to have evolved from parental caregiving behaviour. Empathy requires the ability to form an idea of the emotional state of another. The integrity of the somatosensory cortex appears to be crucial for this characteristic. A study of patients with lesions in this area showed that the affected individuals had considerable difficulty recognising emotions based on facial expressions (Adolphs et al. 2000). It is suspected that this is a disruption of the reconstruction of somatosensory representations, which are apparently necessary to internally perceive or mentally simulate the feeling of the other person. The prefrontal cortex appears to be of paramount importance for empathetic behaviour, as indicated by both neuropsychological tests (Shamay-Tsoory et al. 2003) and experimental investigations into the recognition of emotions based on facial expressions (Hornak et al. 1996). The idea that human empathy for the emotions of others is closely linked to mental simulation is supported by the discovery of so-called mirror neurons. Mirror neurons, discovered in the prefrontal cortex of monkeys, showed activation when an animal observed another performing certain actions (Gallese and Goldman 1999). Similar neurons, which may represent the neural substrate of empathetic feelings, were located in the anterior cingulate gyrus of humans. These neurons responded not only when feeling pain but also when the test subject observed another individual suffering from pain (Hutchison et al. 1999).

From an evolutionary standpoint, the original parental care behavior for offspring has expanded significantly. Through increased connections between limbic structures and the neocortex, this behavior has developed into a broader capacity for care and empathy, extending beyond one's own species to potentially all living beings (MacLean 1985).

3.5 Evolutionary Neuroethology and Psychopathology

The concepts of evolutionary neuroethology provide explanatory models for certain psychopathological phenomena, especially for disorders of affect. The latter can be divided into depressive and manic states. Depressions and manias represent the most common psychiatric entities and are encountered in all human cultures. In terms of their appearance, affective disorders impress as exaggerated forms of physiologically occurring emotional states, so that depression can be seen as an extreme form of the feeling of grief and mania as an exaggerated feeling of elation. Both mental illnesses can occur as isolated (unipolar) disorders or in the form of mixed states, as so-called bipolar disorders (manic-depressive disorders).

Depressive states are primarily characterised by the clinical symptoms of reduced mood and decreased drive. In addition, the patients show loss of appetite, sleep disorders and loss of libido, while pessimistic thoughts predominate. Conversely, in mania there is an increase in mood—which can range from elation to irritability—and drive. The patients feel an inner restlessness and stand out due to their restless behaviour, as well as their flight of ideas.

In addition to the currently held views that these two disease patterns are genetic or neurochemical disorders, there are theories that these states have evolved as adaptive mechanisms during evolution but no longer offer any advantage in the current social context of humans. The starting point for such explanatory models is the ethological principle of dominance behaviour in groups. Ethological studies show that aggressive animal species that live in social groups establish so-called dominance hierarchies to ensure a conflict-free coexistence. These hierarchies are found in all socially organised species, ranging from fish and birds to primates (Schjelderup-Ebbe 1922; de Waal 1986). In such groups, a social ranking is therefore established in which each animal knows its rank in relation to every other group member. This predetermined regulation of socially superior and subordinate group members prevents a new conflict with possible aggressive confrontation every time individuals meet. Comparative studies on primates show that those animals that live in large groups with complex dominance hierarchies have a larger neocortex and thus higher cognitive abilities.

Some authors establish a direct causal relationship between dominance hierarchical social structure and the evolution of the human mind based on this fact (Cummins 1996).

With regard to the emergence of depression and mania in humans, evolutionary theories assume that states of depression, anxiety and excitability represent emotional accompaniments of behaviours necessary for the maintenance of dominance hierarchies in social groups (Price 1967). In this context, a distinction is made between behaviour associated with the maintenance of hierarchies and that associated with changes in the hierarchy. In a functioning hierarchical group, a dominant member reacts to the subordinate with aggressive and irritated behaviour, often accompanied by threatening posture or vocalisation. However, this behaviour can also be quite subtle, such as in the form of a certain gaze. The behaviour of a subordinate group member, on the other hand, is predominantly characterised by fear. Analogous to dominant behaviour, the submissive posture of the subordinate animal can also take discrete forms, such as a brief pause when encountering a superior member. The entire social behaviour of such an organised group is shaped by this hierarchical regulation, so that dominant individuals act rather calmly and confidently in their habitus, while subordinates appear more "nervous" and cautious.

A disruption of the hierarchical structure, on the other hand, leads to instability and poses a danger to the group, especially when the fight between two equally dominant members leads to prolonged disputes. Victory and defeat, and the associated rise or fall in the hierarchy, are naturally associated with corresponding affective states or behavioural patterns. A rise would therefore be associated with a feeling of elation and a fall with a depressive affect. Both reaction patterns would have a stabilising effect on the changed group structure, similar to how aggression and fear contribute to the maintenance of a functioning hierarchy. The evolutionary selection advantage of a depressive behaviour lies primarily in the fact that an individual who has just experienced a defeat is deterred from new combat actions, which is ultimately vital for the individual and the group.

From a comparative ethological perspective, it seems plausible to compare the behaviours of an animal descending in the hierarchy with the symptoms of a depressive patient. Both show signs of fear, social withdrawal and inferiority. Even the noticeable loss of libido and appetite in patients could be explained by the evolutionary purpose of deterring a hierarchically descended animal from attempting to rise again.

From an evolutionary point of view, psychopathological symptoms would thus be seen as so-called excessive residues of behaviours that originally served to stabilise functioning and changed group hierarchies. These excessive residues can be divided into anxious-submissive and aggressive forms. The "excessive anxious-submissive residues" from functioning hierarchies would correspond to the characteristics of neurotic and schizophrenic disorders and the "excessively aggressive residues" to the symptoms of impulsive personality disorders. Analogously, the excessive behavioural residues associated with changes in hierarchies would be compatible with psychiatric disease patterns insofar as the descent would lead to depression and the ascent to mania. Although it is assumed that the selection pressure of these hierarchical behaviours has decreased during human evolution, the persistence of the excessive forms, i.e. the psychopathological symptoms, could be explained by the fact that the point in time when selection ceased is too recent in evolutionary terms to be effective (Price 1967).

The lack of a valid pathophysiological concept of manic-depressive disorders also prompted other authors to support the theories presented so far and to maintain an evolutionary theoretical model of these entities. In a psychiatric study, an attempt was made to demonstrate a congruence between social rank and associated behaviours or character traits (Gardner 1982). As part of this biographical investigation of selected well-known leadership personalities, Gardner postulated that these high-ranking individuals exhibited a strikingly high degree of energy, enjoyment of work, sociability and humour, but also a certain restlessness and irritability (Gardner 1982). These results would support the previously described theories about dominance hierarchies insofar as a correlation between rank and associated behaviour is demonstrated here. Furthermore, it can be inferred from this that pathological conditions could possibly be just an (extreme) variant of normal behaviour.

The concept of dominance hierarchies was supplemented by the introduction of a term from behavioural biology, namely the "Resource-Holding Potential" (RHP). This refers to animals' knowledge of their own fighting capacity. Such an animal "self-concept" possibly represents the evolutionary precursor of human self-esteem (Parker 1974) and determines whether an animal chooses the escalation strategy and attacks in a confrontation, or tends to deescalate and behaves submissively, in the sense of an involuntary subordinate strategy (ISS). The behaviour of subordination could be seen in humans as a precursor to a depressive disorder.

Since human social behaviour cannot directly be spoken of dominance hierarchies, some authors subsequently introduced the term "Social Competition", which takes into account the phylogenetic change of human social hierarchies and their associated behaviours (Price et al. 1994). The Social Competition Hypothesis states, in reference to the dominance hierarchy concept and with regard to the emergence of depression, that humans, like group-living animals, have the ability to give in in competitive situations. This subordination behaviour prevents a new confrontation through the associated feeling of weakness and inability, on the one hand, and signals a sign of surrender to the opponent through this behaviour, on the other hand. Ultimately, the feeling of surrender leads to an acceptance of the conflict outcome. In the case of an impossibility or disturbance of voluntary surrender, however, there is an abnormal and prolonged subordination behaviour, which eventually leads to a depressive disorder.

The evolutionary explanatory model of psychopathological phenomena can be significantly supplemented by Paul MacLean's evolutionary neuroethological concept (MacLean 1990). The possible behavioural strategies of fighting (escalation) and subordination (deescalation) postulated within the framework of dominance hierarchies or the "Social Competence Hypothesis" could be viewed from the perspective of the evolutionary-hierarchical brain model (Triune Brain). The three neural associations of the forebrain, which can be simplified into rational part (neocortex), emotional part (limbic system) and instinctive part (striatum), are indeed interconnected but function largely independently of each other. At the level of the neocortex, a rational decision is made in a conflict either for a confrontation or for subordination. At the level of the limbic system, the corresponding emotions are generated depending on the decision, i.e. anger, aggression and arousal in the case of escalation, as well as feelings of fear, depression and despair in the context of deescalation. At the level of the striatum (reptilian brain or R-complex), according to this theory, there are corresponding changes in mood, i.e. an elevation of mood during escalation and a reduction during deescalation (Price 2002). However, the three neural associations differ not only in terms of their specific functions but also in terms of the degree of consciousness. While the neocortical function is equivalent to conscious and voluntary action, the limbic system has only limited access to consciousness and thus also limited knowledge of the cause of its emotional decisions or actions. The decisions generated by the striatum or R-complex occur, according to their instinctive nature, entirely involuntarily. A crucial aspect of affective disorders is the fact that states of depression or mania cannot be influenced by cognitive processes. The question of why the evolution of the neocortex did not lead to complete control of the subcortical emotional and instinctive centres could be answered by attributing an adaptive function to the emotions and behaviours associated with conflict situations. Based on the Triune Brain Theory, in the case of a social conflict situation, each of the three associations of the forebrain has different action options with regard to an escalation or deescalation of the conflict (Cory and Gardner 2002).

The solution approaches at the rational or neocortical level are insofar as evident, when an individual classifies the contractor as superior and rationally recognises themselves as the inferior. However, this most favourable outcome of a conflict situation is only given when the cognitive decision correlates with the emotional and instinctive tendencies of subcortical structures.

The functional and evolutionary significance of the subcortical structures for problem-solving is illustrated by the example of another possible conflict constellation: In the case of a discrepancy between the rational neocortical decision to fight (escalation) and the emotional or instinctive decision to give up (deescalation), the subcortical processes allow an influence on the cognitive strategy, insofar as the original goal no longer appears desirable and the individual feels unsuitable or unworthy. The instinctive-emotional influence thus causes a pessimistic colouring of the original positive goal setting and ultimately leads in this favourable case to an insightful surrender and acceptance of the loser role. The hierarchically subordinate brain structures gain the upper hand in this decision and ultimately determine the outcome of the conflict. This instinctive-emotional "gut feeling" seems to represent the warning function for the escalation-prone neocortical system in evolution (Gardner and Price 1999).

From the perspective of the Triune Brain Theory and the Social Competition Hypothesis, more differentiated aspects also emerge with regard to the genesis of depression. According to this, depression would be caused by prolonged instinctive deescalation, which in turn is caused by prolonged and extreme rational escalation. A prolonged instinctive deescalation can, according to this model, also be induced by an inappropriate emotional escalation. This mechanism would be effective in individuals after traumatic events, when an event is rationally accepted, but an emotional escalation in the form of anger or rage persists, which ends in depression after a decision for instinctive deescalation. Another variant of the conflict constellations is seen in patients who emotionally deescalate and rationally escalate, for example by presenting themselves as emotionally suffering, but cognitively pursuing goals that lead to these emotional problems. An individual can also deescalate on all levels. The associated submission can then be accepted or rejected by the opponent or partner. Finally, there can also be an abnormal sensitivity and abnormal tendency to deescalate at the instinctive level, associated with an increased sensitivity of the affected individuals (Price 2002).

The attempt to see psychopathological phenomena as manifestations of archaic evolutionary adaptation mechanisms opens up a new dimension in understanding disorders of the human mind (Stevens and Price 2000). This research area, also referred to as evolutionary psychiatry, draws on comparative ethological and neurobiological data as an explanatory model for psychopathology and also offers the opportunity to integrate other theories of the mind and psyche. One of these theories, which significantly influenced the self-understanding of humans, is psychoanalysis.

Bibliography

Adolphs R, Damasio H, Tranel D et al. (2000) A role for somatosensory cortices in the visual recognition of emotion as revealed by 3-D lesion mapping. J. Neurosci. 20: 2683–2690.

Adolphs R, Tranel D, Damasio H, Damasio A (1994) Impaired recognition of emotion in facial expressions following bilateral damage to the human amygdala. Nature 372: 669–672.

Adolphs R, Tranel D, Hamann S et al. (1999) Recognition of facial emotion in nine subjects with bilateral amygdala damage. Neuropsychologia 37: 1111–1117.

Albe-Fessard D, Buser P (1979) Is a theory on neostriatal functions now possible?. In: Divac I and Oberg RGE (Eds.) The Neostriatum, Pergamon Press, Elmsford, N.Y., 315–319.

Anschel S (1977) Functional specificity of vocalizations elicited by electrical brain stimulation in the turkey. Brain Behav. Evol. 14: 399–417.

Akert K, Gruesen RA, Woolsey CN, Meyer DR (1961) Klüver-Bucy syndrome in monkeys with neocortical ablations of temporal lobe. Brain 84: 480–498.

Arroyo S, Lesser RP, Gordon B, Uematsu S, Hart J, Schwerdt P et al. (1993) Mirth, laughter and gelastic seizures. Brain 116: 757–780.

Auffenberg W (1978) Social and feeding behaviour in Varanus komodoensis. In: Greenberg N, MacLean PD, (Eds.) The behaviour and neurology of lizards. U.S. Government Printing Office, Washington, D.C., pp 301–331.

Aull-Watschinger S, Pataraia E, Baumgartner C (2008) Sexual auras: predominance of epileptic activity within the mesial temporal lobe. Epilepsy Behav. 12: 124–127.

Auffenberg W (1981) The behavioural ecology of the Komodo Monitor. University Presses of Florida, Gainesville.

Baerends GP (1956) Aufbau des tierischen Verhaltens. In: Helmcke JG (Ed) Handbuch der Zoologie, Band VIII, de Gruyter, Berlin.

Baumgartner C, Gröppel G, Leutmezer F, Aull-Watschinger S, Pataraia E, Feucht M, Trinka E, Unterberger I, Bauer G (2000) Ictal urinary urge indicates seizure onset in the non-dominant temporal lobe. Neurology 55: 432–434.

Beach FA (1951) Effects of forebrain injury upon mating behaviour in male pigeons. Behaviour 4: 36–59.

Bellairs A (1970) The life of reptiles. Universe Books, New York.

Buchanan TW, Tranel D, Adolphs R (2005). Emotional autobiographical memories in amnesic patients with medial temporal lobe damage. J. Neurosci. 25: 3151–3160.

Burgess PW, Shallice T (1996) Bizarre responses, rule detection and frontal lobe lesions. Cortex 32: 241–259.

Brown JW (1967) Physiology and phylogenesis of emotional expression. Brain Res. 5: 1–14.

Buzsaki G (2002) Theta oscillations in the hippocampus. Neuron 33: 325–340.

Caine ED, Hunt RD, Weingartner H, Ebert MH (1978) Huntington's dementia. Clinical and neuropsychological features. Arch. Gen. Psychiatry 35: 377–384.

Calamandrei G, Keverne EB (1994) Differential expression of Fos protein in the brain of female mice dependent on pup sensory cues and maternal experience. Behav. Neurosci. 108: 113–120.

Camara Magalhaes S, Vitorino Souza C, Rocha Dias T, Felipe Carvalhedo de Bruin P, de Bruin VM (2005) Lifestyle regularity measured by the social rhythm metric in Parkinson's disease. Chronobiol Int. 22: 917–924.

Cascino GD, Sutula TP (1989) Thirst and compulsive water drinking in medial basal limbic epilepsy: an electroclinical and neuropathological correlation. J. Neurol. Neurosurg. Psychiatry 52: 680–681.

Ceccaldi M, Milandre L (1994) A transient fit of laughter as the inaugural symptom of capsular-thalamic infarction. Neurology 44: 1762.

Colbern D, Isaacson RL, Bohus B, Gispen WH (1977) Limbic-midbrain lesions and ACTH-induced excessive grooming. Life Sci. 21: 393–402.

Cory GA, Gardner R (2002) The evolutionary neuroethology of Paul MacLean. Convergences and Frontiers. Praeger Publishers, Westport, CT.

Creutzfeld OD (1983) Cortex Cerebri. Springer, Berlin.

Cummins DD (1996) Dominance hierarchies and the evolution of human reasoning. Minds and Machines 6: 463–480.

Damasio AR, Tranel D, Damasio H (1991) Somatic markers and the guidance of behaviour. In: Levin H, Eisenberg H, Benton A (Eds) Frontal lobe function and dysfunction. Oxford Univ. Press, New York, pp 217–228.

de Waal FBM (1986) The integration of dominance and social bonding in primates. Q. Rev. Biol. 61: 459–479.

Dobesberger J, Walser G, Unterberger I, Embacher N, Luef G, Bauer G, Benke T, Bartha L, Ulmer H, Ortler M, Trinka E (2004) Genital automatisms: a video-EEG study in patients with medically refractory seizures. Epilepsia 45: 777–780.

Eibl-Eibesfeldt I (1971) !Ko-Buschleute (Kalahari) – Schamweisen und Spotten. Homo 22: 261–266.

Eibl-Eibesfeldt I (1984) Die Biologie des menschlichen Verhaltens. Grundriß der Humanethologie. Piper, München/Zürich.

Eibl-Eibesfeldt I (1999) Grundriß der vergleichenden Verhaltensforschung. Piper, München.

Enard W, Przeworski M, Fisher SE, Lai CSL, Wiebe V, Kitano T, Monaco AP, Paäbo S (2002) Molecular evolution of FOXP2, a gene involved in speech and language. Nature 418: 869–872.

Engel J, Jr. (1996) Introduction to the temporal lobe epilepsies. Epilepsy Res. 26: 141–150.

Engel JrJ, Williamson PD, Wieser HG (1997) Mesial temporal lobe epilepsy. In: Engel JrJ, Pedley TA (Eds) Epilepsy: A comprehensive textbook. Philadelphia: Lippincott-Raven Publishers.

Fish DR, Gloor P, Quesney FL, Olivier A (1993) Clinical responses to electrical brain stimulation of the temporal and frontal lobes in patients with epilepsy. Pathophysiological implications. Brain 116: 397–414.

Fleischer S, Slotnick BM (1978) Disruption of maternal behaviour in rats with lesions of the septal area. Physiol. Behav. 21: 189–200.
Fried I, Wilson CL, MacDonald KA, Behnke EJ (1998) Electric current stimulates laughter. Nature 391: 650.
Gajdusek DC (1970) Physiological and psychological characteristics of Stone Age man. In: Symposium on biological bases of human behaviour. Eng. Sci. 33, 26–33, 56–62.
Gallese V, Goldman A (1999) Mirror neurons and the simulation theory of mind-reading. Trends Cogn. Sci. 2: 493–500.
Gardner H, Ling PK, Flamm L, Silverman J (1975) Comprehension and appreciation of humorous material following brain damage. Brain 98: 399–412.
Gardner R (1982) Mechanisms in manic-depressive disorder. An evolutionary model. Arch. Gen. Psychiatry 39: 1436–1441.
Gardner R, Price JS (1999) Sociophysiology and depression. In: Joiner T, Coyne JC (Eds) The interactional nature of depression: Advances in interpersonal approaches, APA Books, Washington DC, pp 247–268.
Goel V, Dolan RJ (2001) The functional anatomy of humour: segregating cognitive and affective components. Nature Neurosci. 4: 237–238.
Goldby F, Gamble HJ (1957) The reptilian cerebral hemispheres. Biol Rev. 32: 383–420.
Gould SJ (1977) Ontogeny and Phylogeny. The Belknap Press of Harvard University Press, Cambridge, MA; London, England.
Graybiel AM (2008) Habits, rituals, and the evaluative brain. Annu. Rev. Neurosci. 31: 359–387.
Greenberg N, Ferguson JL, MacLean PD (1976) A neuroethological study of display behaviour in lizards. Neuroscience 2: 689.
Greenberg N (1978) Ethological considerations in the experimental study of lizard behaviour. In: Greenberg N, MacLean PD, (Eds.) The behaviour and neurology of lizards. U.S. Government Printing Office, Washington, D.C., pp 203–224.
Greenberg N, MacLean PD, Ferguson JL (1979) Role of the paleostriatum in species-typical behaviour of the lizard Anolis Carolinensis. Brain Res. 172: 229–241.
Halgren E, Walter RD, Cherlow DG, Crandall PH (1978) Mental phenomena evoked by electrical stimulation of the human hippocampal formation and amygdala. Brain 101: 83–117.
Harris VA (1964) The life of the Rainbow Lizard. Hutchinson, London.
Hassler R (1961) Motorische und sensible Effekte umschriebener Reizungen und Ausschaltungen im menschlichen Zwischenhirn. Dtsch. Z. Nervenheilkd. 183: 148–171.
Heath RG (1963) Electrical self-stimulation of the brain in man. Am. J. Psychiatry 120: 571–577.

Hebben N, Corkin S, Eichenbaum H, Shedlack K (1985) Diminished ability to interpret and report internal states after bilateral medial temporal resection: case H.M. Behav. Neurosci. 99: 1031–1039.

Hornak JE, Rolls T, Wade T (1996) Face and voice expression identification in patients with emotional and behavioural changes following ventral frontal lobe damage. Neuropsychologia 34: 247–261.

Hutchison WD, Davis KD, Lozano AM et al. (1999) Pain-related neurons in the human cingulate cortex. Nat. Neurosci. 2: 403–405.

Jansky J, Ebner A, Szupera Z, Schulz R, Hollo A, Szücs A, Clemens B (2004) Orgasmic aura: a report of seven cases. Seizure 13: 441–444.

Jerison HJ (1973) Evolution of the brain and intelligence. Academic Press, New York and London.

Kallen B (1951) On the ontogeny of the reptilian forebrain. Nuclear structures and ventricular sulci. J Comp. Neurol. 95: 307–347.

Karnath HO, Thier P (2003) Neuropsychologie. Springer Verlag, Berlin, Heidelberg, New York.

Karnath HO, Wallesch CW, Zimmermann P (1991) Mental planning and anticipatory processes with acute and chronic frontal lobe lesions: a comparison of maze performance in routine and non-routine situations. Neuropsychologia 29: 271–290.

Kataoka S, Hori A, Shirakawa T, Hirose G (1997) Paramedian pontine infarction. Neurological/topographical correlation. Stroke 28: 809–815.

Kim JS, Choi-Kwon KS (2000) Poststroke depression and emotional incontinence: correlation with lesion location. Neurology 54: 1805–1810.

Krolak-Salmon P, Henaff MA, Vighetto A, Bauchet F, Bertrand O, Maugiere F et al (2006) Experiencing and detecting happiness in humans: The role of the supplementary motor area. Ann. Neurol. 59: 196–199.

Leutmezer F, Serles W, Bacher J, Gröppel G, Pataraia E, Aull S, Olbrich A, Czech T, Baumgartner C (1999) Genital automatisms in complex partial seizures. Neurology 52: 1188–1191

Loddenkemper T, Foldvary N, Raja S, Neme S, Lüders HO (2003) Ictal urinary urge: further evidence for lateralisation to the non-dominant hemisphere. Epilepsia 44: 124–126.

Lorenz K (1978) Vergleichende Verhaltensforschung. Grundlagen der Ethologie. Springer, Wien.

Luciano D, Devinsky O, Perrine K (1993) Crying seizures. Neurology 43: 2113–2117.

MacLean PD (1952) Some psychiatric implications of physiological studies on frontotemporal portion of limbic system (visceral brain). Electroencephalogr. Clin. Neurophysiol. 4: 407–418.

MacLean PD (1954) The limbic system and its hippocampal formation. Studies in animals and their possible application to man. J Neurosurg. 11: 29–44.

MacLean PD (1957) Visceral functions of the nervous system. Annu. Rev. Physiol. 19: 397–416.
MacLean PD (1962) New findings relevant to the evolution of psychosexual functions of the brain. J. Nerv. Ment. Dis. 135: 289–301.
MacLean PD (1964) Mirror display in the squirrel monkey *Saimiri sciureus*. Science 146: 950–952.
MacLean PD (1969) The internal-external bonds of the memory process. J. Nerv. Ment. Dis. 149: 40–47.
MacLean PD (1970) The triune brain, emotion, and scientific bias. In: Schmitt FO (Ed) The Neurosciences. Second study program. Rockefeller University Press, New York, 336–349.
MacLean PD (1973) The brain's generation gap: Some human implications. Zygon J. Relig. Sci. 8, 113–127.
MacLean PD (1975a) On the evolution of three mentalities. Man-Environment-Systems 5: 213–224.
MacLean PD (1975b) The imitative-creative interplay of our three mentalities. In: Harris H (Ed.) Astride the two cultures: Arthur Koestler at 70, Hutchinson, London, pp 187–211.
MacLean PD (1975c) An ongoing analysis of hippocampal inputs and outputs: Microelectrode and anatomic findings in squirrel monkeys. In: Isaacson RI, Pribram KH (Eds) The Hippocampus, Volume 1, Plenum Press, New York, pp 177–211.
MacLean PD (1977) An evolutionary approach to brain research on prosematic (nonverbal) behaviour. In: Rosenblatt JS and Komisaruk BR (Eds) Reproductive Behaviour and Evolution. Plenum Press, New York, pp 177–211.
MacLean PD (1978) Effects of lesions of globus pallidus on species-typical display behaviour of squirrel monkeys. Brain Res. 149: 175–196.
MacLean PD (1985) Brain evolution relating to family, play, and the separation call. Arch. Gen. Psychiatry 42: 405–417.
MacLean PD (1987) The midline frontolimbic cortex and the evolution of crying and laughter. In: Perecman E (Ed) The frontal lobes revisited. IRBN Press, New York, pp 121–140.
MacLean PD, Newman JD (1988) Role of midline frontolimbic cortex in production of the isolation call of squirrel monkeys. Brain Res. 450: 111–123.
MacLean PD (1990) The Triune Brain in evolution. Role in paleocerebral functions. Plenum Press, New-York, London.
Marín O, Smeets WJAJ, Gonzáles A (1998) Basal ganglia organisation in amphibians: evidence for a common pattern in tetrapods. Prog. Neurobiol. 55: 363–397.
McGaugh JL (2004) The amygdala modulates the consolidation of memories of emotionally arousing experiences. Annu. Rev. Neurosci. 27: 1–28.
Mellars P (1998) Neanderthals, modern humans and the archaeological evidence for language. In: Jablonski NG, Aiello LC (Eds) The origin and diversification of language. Academy of Sciences, San Francisco.

Mendez MF, Nakatawase TV, Brown CV (1999) Involuntary laughter and inappropriate hilarity. J. Neuropsychiatry Clin. Neurosci. 11: 253–258.

Montagu A (1959) Natural selection and the origin and evolution of weeping in man. Science 130: 1572–1573.

Morgane PJ, Galler JR, Mokler DJ (2005) A review of systems and networks of the limbic forebrain/limbic midbrain. Progr. Neurobiol. 75: 143–160.

Murphy MR, MacLean PD, Hamilton SC (1981) Species-typical behaviour of hamsters deprived from birth of neocortex. Science 213: 459–461.

Mullan S, Penfield W (1959) Illusions of comparative interpretation and emotion; production by epileptic discharge and by electrical stimulation in the temporal cortex. AMA Arch. Neurol. Psychiatry 81: 269–284.

Penfield W, Welch K (1951) The supplementary motor area of the cerebral cortex. Arch. Neurol. Psychiatry 66: 289–317.

Newman JD (1985) The infant cry of primates. An evolutionary perspective. In: Lester BM, Boukydis CFZ (Eds), Infant Crying: Theoretical and Research Perspectives. Plenum Press, New York.

Nieuwenhuys R, Voogd J, van Huijzen C (2007) The human central nervous system. 4th Ed. Springer.

Noble GK, Bradley HT (1933) The mating behaviour of lizards: its bearing on the theory of sexual selection. Ann N.Y Acad. Sci. 35: 25–100.

Owren MJ, Bachorowski JA (2003) Reconsidering the evolution of non-linguistic communication: the case of laughter. J. Nonverbal Behav. 27: 183–197.

Panksepp J (1998) Affective Neuroscience: The foundations of human and animal emotions. Oxford University Press, New York and Oxford.

Panksepp J (2005) Beyond a joke: from animal laughter to human joy. Science 308: 62–63.

Parker GA (1974) Assessment strategy and the evolution of fighting behaviour. J. Theoret. Biol. 47: 223–243.

Parvizi J, Anderson SW, Martin CO, Damasio H, Damasio AR (2001) Pathological laughter and crying. A link to the cerebellum. Brain 124: 1708–1719.

Ploog DW, MacLean PD (1963) Display of penile erection in squirrel monkey (*Saimiri sciureus*). Anim. Behav. 11: 32–39.

Plotnik R (1968) Changes in social behaviour of squirrel monkeys after anterior temporal lobectomy. J. Comp. Physiol. Psychol. 66: 369–377.

Poeck K, Pilleri G (1963) Pathologisches Lachen und Weinen. Schweiz. Arch. Neurol. Neurochir. 92: 323–370.

Poeck K (1969) Pathophysiology of emotional disorders associated with brain damage. In: Vinkens PJ, Bruyn GW (Eds) Handbook of clinical neurology. Elsevier, Amsterdam pp 343–367.

Poeck K (1985) Pathological laughter and crying. In: Vinken PJ, Bruyn GW, Klawans HL (Eds) Handbook of clinical neurology, vol 1: Clinical neuropsychology. Elsevier, Amsterdam pp 257–263.

Pontius AA (1993) Neuroethological aspects of certain limbic seizure-like dysfunctions: Exemplified by limbic psychotic trigger reaction (motiveless homicide with intact memory). Integrative Psychiatry 9: 151–167.

Pontius AA (2002) Neuroethology, exemplified by limbic seizures with motiveless homicide in "Limbic Psychotic Trigger Reaction". In: Cory GA, Gardner R (Eds) The evolutionary neuroethology of Paul MacLean. Convergences and Frontiers. Praeger Publishers, Westport, CT, pp 167–191.

Preuss TM (2004) What is it like to be a human? In: Gazzaniga MS (Ed) The cognitive neurosciences III. The MIT Press, Cambridge, London, pp 5–22.

Price J (1967) The dominance hierarchy and the evolution of mental illness. Lancet 2: 243–246.

Price J (2002) The triune brain, escalation, de-escalation strategies, and mood disorders. In: Cory GA, Gardner R (Eds) The evolutionary neuroethology of Paul MacLean. Convergences and Frontiers. Praeger Publishers, Westport, CT, pp 107–117.

Price J, Sloman L, Gardner R, Gilbert P, Rohde P (1994) The social competition hypothesis of depression. British J. Psychiatry 164: 309–315.

Ramachandran VS (1998) The neurology and evolution of humour, laughter and smiling: The false alarm theory. Med. Hypotheses 51: 351–354.

Ramachandran VS (2003) Humour and laughter: A biological hypothesis. In: Baars BJ, Banks WP, Newman JB (Eds) Essential sources in the scientific study of consciousness. The MIT Press, Cambridge, London, pp 819–820.

Rolls ET, Thorpe SJ, Maddison S, Roper-Hall A, Puerto A, Perret D (1979) Activity of neurons in the neostriatum and related structures in the alert animal. In: Divac I and Oberg RGE (Eds) The Neostriatum. Pergamon Press, Elmsford, N.Y., pp 163–182.

Saka E, Goodrich C, Harlan P, Madras BK, Graybiel AM (2004) Repetitive behaviours in monkeys are linked to specific striatal activation patterns. J. Neurosci. 24: 7557–7565.

Salk L (1960) The effects of the normal heartbeat sound on the behaviour of the newborn infant: Implications for mental health. World Ment. Health 12: 168–175.

Schjelderup-Ebbe (1922) Beiträge zur Sozialpsychologie des Haushuhns. Zeitschrift für Psychologie 88: 225–252.

Seiss R (1965) Beobachtungen zur Frage der Übersprungsbewegungen im menschlichen Verhalten. Psychol. Beitr. 8: 1–97.

Shamay-Tsoory SG, Tomer R, Berger BD, Aharon-Peretz J (2003) Characterisation of empathy deficits following prefrontal brain damage: The role of the right ventromedial prefrontal cortex. J. Cogn. Neurosci. 15: 324–337.

Shammi P, Stuss DT (1999) Humour appreciation: a role of the right frontal lobe. Brain 122: 657–666.

Singer HS, Reiss AL, Brown JE et al. (1993) Volumetric MRI changes in basal ganglia of children with Tourette syndrome. Neurology 43: 950–956.

Slotnick BM, Nigrosh BJ (1975) Maternal behaviour in mice with cingulate cortical, amygdala, or septal lesions. J. Comp. Physiol. Psychol. 88: 118–127.

Spitz RA (1965) Vom Säugling zum Kleinkind (11.Auflage, 1987). Klett-Cotta, Stuttgart.

Stevens A, Price J (2000) Evolutionary Psychiatry: A new beginning (Second Edition). Routledge, London, New York.

Striedter GF (2005) Principles of brain evolution. Sinauer Associates, Sunderland, MA.

Stroufe LA, Waters E (1976) The ontogenesis of smiling and laughter: A perspective on the organisation of development in infancy. Psych. Rev. 83: 173–189.

Swanson LW (1987) Limbic system. In: Adelman G. (Ed.). Encyclopedia of Neuroscience. Birkhäuser, Boston, Basel, Stuttgart, 589–591.

Taylor AE, Saint-Cyr JA (1995) The neuropsychology of Parkinson's disease. Brain Cogn. 28: 281–296.

Tinbergen N (1940) Die Übersprungsbewegung. Z. Tierpsychol. 4: 1–40.

Tinbergen N (1951) The study of instinct. Oxford University Press, London.

Trinka E, Walser G, Unterberger I, Luef G, Benke T, Bartha L, Ortler M, Bauer G (2003) Peri-ictal water drinking lateralizes seizure onset to the nondominant temporal lobe. Neurology 60: 873–876.

Urbach E, Wiethe C (1929) Lipoidosis cutis et mucosae. Virchows Arch. 273: 285–319.

Van Buren JM, Li CL, Ojemann GA (1966) The fronto-striatal arrest response in man. Electroencephalogr. Clin Neurophysiol. 21: 114–130.

Vogt C, Vogt O (1920) Lehre der Erkrankungen des striären Systems. J. Psychol. Neurol 25: 628–846.

Weiss P (1941) Self-differentiation of the basic patterns of coordination. Comp. Psychol. Monogr. 17: 1–96.

von Wickler W (1966) Ursprung und biologische Deutung des Genitalpräsentierens männlicher Primaten. Z. Tierpsychol. 23: 422–437.

Wieser HG (1983) Electroclinical features of the psychomotor seizure. Gustav Fischer, Stuttgart- New York.

Wiest G, Lehner-Baumgartner E, Baumgartner C (2006) Panic attacks in an individual with bilateral selective lesions of the amygdala. Arch. Neurol. 63: 1798–1801.

Wild B, Rodden FA, Grodd W, Ruch W (2003) Neural correlates of laughter and humour. Brain 126: 2121–2138.

Williams D (1956) The structure of emotions reflected in epileptic experiences. Brain 79: 29–67.

Witt K, Nuhsman A, Deuschl G (2002). Dissociation of habit-learning in Parkinson's and cerebellar disease. J. Cogn. Neurosci. 14: 493–499.

Wojtecki L, Nickel J, Timmermann L, Maarouf M, Südmeyer M et al. (2007) Pathological crying induced by deep brain stimulation. Mov. Disord. 22: 1314–1316.

Yakovlev PI (1948) Motility, behaviour, and the brain. Stereodynamic organisation and neural coordinates of behaviour. J. Nerv. Ment. Dis. 107: 313–335.

Yin HH, Knowlton BJ (2006) The role of the basal ganglia in habit formation. Nature Rev. Neurosci 7: 464–476.

Zeilig G, Drubach M, Katz-Zeilig M, Karatinos J (1996) Pathological laughter and crying in patients with closed traumatic brain injury. Brain Inj. 10: 591–597.

CHAPTER 4

Hierarchical Concepts in Psychoanalysis

Starting from observations on neurotic patients, psychoanalysis was developed as a method for treating psychopathological disorders. Thus, it not only represents a treatment technique with its own theory of disease but also has its own research method, which allows mental processes to be investigated and the collected data to be integrated into its own theory of the human psyche, the psychoanalytic metapsychology. Not least because of its approach of placing the subject and its relation to inner and outer reality at the centre of its interest, psychoanalysis is often still considered the most coherent and comprehensive theory of the mind (Kandel 2006). The focus of psychoanalysis on the subjective self connects it in certain aspects with the evolutionary neuroethology of Paul MacLean, who described this research approach with the term "epistemics" (MacLean 1975).

In metapsychology, the insights gained from clinical experience are anchored in the form of basic concepts or basic assumptions of psychoanalysis. These enable every form of mental activity (from dreams to fantasies or ideas) and every mental conflict (from neurotic to psychotic and personality disorders) to be viewed from certain theoretical perspectives, i.e. from a dynamic, economic, genetic, adaptive and structural aspect. The dynamic aspect refers to the psychological forces at work in a conflict (such as in drives or affects), with the economic aspect directly related to these forces and referring to the associated psychoenergetic balance. The

G. Wiest, *Hierarchies in the Brain, Mind and Behaviour*, https://doi.org/10.1007/978-3-031-78030-1_4

genetic viewpoint assumes that all mental phenomena can be assessed in terms of origin, maturation and development. The adaptive aspect refers to the assumption that the development of the mental apparatus requires ongoing adaptation between inner and outer reality. The structural viewpoint finally assumes that all mental phenomena can be understood with reference to a structural organisation of the mental apparatus and the inherent mechanisms (Loch 1999).

Influenced by the theories of Darwin, Spencer and Hughlings Jackson, Freud adopted the concept of the hierarchically organised nervous system and made it, among other theoretical elements, the basis for a model of the mental apparatus and the human psyche, although he—not least due to a lack of neurobiological data—did not seek topical correlations between brain structure and mental function.

Hierarchical models or principles are found in psychoanalysis mainly in the following important basic concepts: (1) In the metapsychological models of the mental apparatus; (2) in the basic principles of mental functioning, the primary and secondary process; (3) in the concept of regression and (4) in the phylogenetic and ontogenetic aspects of psychoanalytic theory.

4.1 Psychoanalytic Models of the Mental Apparatus

Understanding psychoanalytic theory is sometimes made difficult by the fact that seemingly different models exist side by side for certain concepts. This also applies to the models of the human mind, or the so-called mental or psychic apparatus, proposed by Freud. The reason for the coexistence of different schemas is the fact that every psychoanalytic theory development must be seen in the context of the existing state of knowledge at that time. Thus, the discovery of the importance of drives for the individual led to a departure from the original seduction theory—according to which neurotic disorders are due to sexual seduction in childhood—and to the development of the first concept of the mental apparatus, the topographical model.

The Topographical Model of the Mental Apparatus

The topographical model of the mental apparatus was presented by Freud in its basic structure in the *Interpretation of Dreams* (Freud 1900a). The conception of this model thus coincides, on the one hand, with the discovery of the significance and mechanisms of dreams and, on the other hand, with the formulation of the psychoanalytic drive theory, which emphasised the importance of the drive forces for the mental development of the individual and in the genesis of neuroses. The concept of the drive, which was initially conceived only as libidinous or sexual, is special in that it must be presented in the border area between soma and psyche. In this sense, the drive is a "psychic representation of a somatic source of stimulation". The insights of psychoanalysis into the functioning of the human mind, which are predominantly based on the study of mental conflicts, imply a model of the psyche with dynamic interaction of different instances. In this context, the term "topographical model" of the mental apparatus can be understood as an attempt to conceive a topography of the mind that takes into account the different structures of the psyche. The central aspect on which the first models of the mental apparatus are based is the assessment of the degree of consciousness of certain mental contents.

As a precursor to the topographical model, Freud designed a simple scheme that corresponds to the mental apparatus of the unstructured psyche. This model is based on his postulated "Principle of constancy", according to which the nervous system always strives to keep the sum of excitations constant. The mental apparatus could—according to this principle—protect itself from overstimulation by reacting with immediate motor discharge. This model is unmistakably based on the neurological "reflex arc model", in which a stimulation of the muscle spindles leads to motor reflex movement. In the mental apparatus, therefore, there is a system at the sensitive end that receives the perceptions (W), while at the other end the motor discharge (M) is achieved through access to motility (Fig. 4.1). According to these ideas, every mental activity starts from internal or external stimuli and ends in innervation, which determines the direction of the process (Freud 1900a).

At the beginning of mental development, the infant seeks to end the excitation or stimulation caused by internal needs in the shortest possible way. Since direct satisfaction is denied to him due to his limited motor possibilities, he tries to satisfy the persistent stimulation through a "hallucinatory" wishful thinking. However, since this is not a real satisfaction,

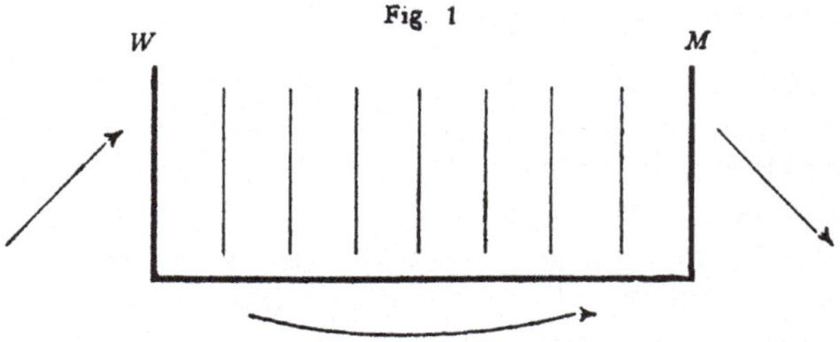

Fig. 4.1 Freud's first scheme of the mental apparatus, which is based on the neurological "reflex arc model". W = Perception, M = Motor skills (Freud 1900a)

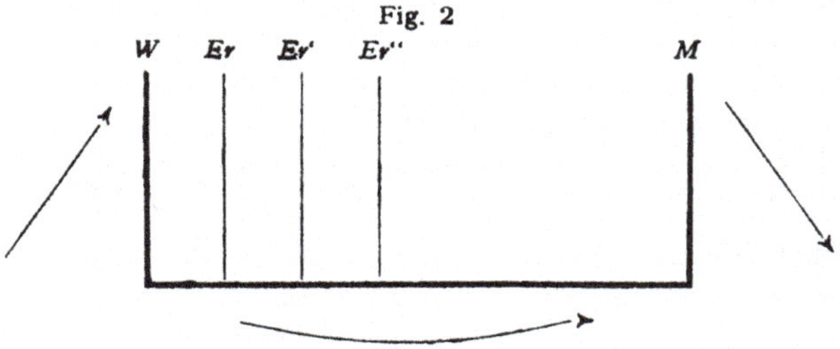

Fig. 4.2 Freud's model of the psychic apparatus, supplemented by the function of memory (memory trace). W = Perception, M = Motor skills, Er, Er′, Er″= Memory traces (Freud 1900a)

a second mental system needs to be developed that controls the external reality through the ability to voluntary motor control and guarantees real satisfaction. In later life, only in dreams and in psychosis is there a return to this infantile psychic function, in the form of hallucinatory wish fulfilment. In this possibility of returning to earlier psychic organisational forms, in the form of a "regression", there is another parallel to the hierarchical principle of the nervous system by Hughlings Jackson (see chapter on regression).

Freud subsequently supplements this model of the unstructured psyche with the concept of "memory traces" (Er). These functions correspond to memory and are an expression of a more differentiated psychic apparatus, which not only discharges perceptions or stimuli but can also store their contents (Fig. 4.2).

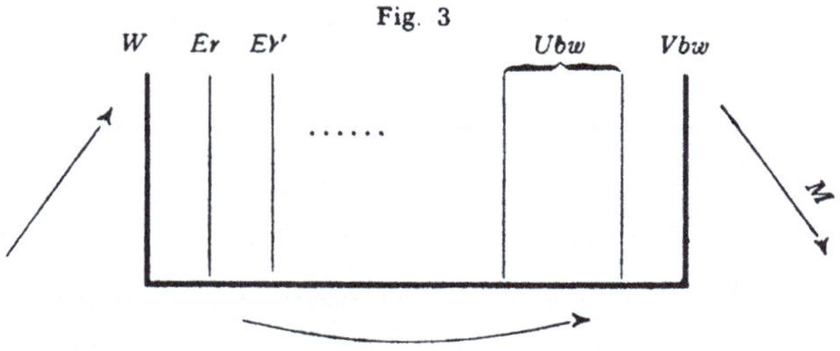

Fig. 4.3 Freud's model of the psychic apparatus, supplemented by the function of memory (memory trace) and the systems Preconscious and Unconscious. W = Perception, M = Motor skills, Er, Er′, Er″ = Memory traces, Ubw = System Unconscious, Vbw = System Preconscious (Freud 1900a)

The study of dream mechanisms prompted Freud to further modify the model by expanding it with the systems Preconscious (Vbw) and Unconscious (Ubw) (Fig. 4.3).

The Vbw system is located near the motor end, as the associated excitation processes can reach consciousness without a barrier. The Ubw system, on the other hand, is located behind the Vbw system, expressing that it has no direct access to consciousness except through the Preconscious (Freud 1900a).

The stimulation or excitation of the psychic apparatus can not only originate from the sensitive end but can also be caused by an excitation from the Unconscious, such as by instinctual impulses. Since dreams are induced by preconscious day residues (thoughts, affects from daily experiences), the excitation can also originate from the Preconscious. Ideally, however, the excitation of the psychic apparatus begins with a sensory stimulation and passes through subsequently the systems Ubw, Vbw and Conscious, to ultimately end in a motor discharge. This scheme already leads to the actual topographical model of the psychic apparatus. The central aspect under which the topographical system was developed is the quality of consciousness. Accordingly, the individual systems—the systems Unconscious, Preconscious and Conscious—have a hierarchical topography that correlates with their degree of consciousness from depth to surface (Fig. 4.4).

The systems are to be seen as a kind of continuum among each other, but in the event of a psychic conflict, there can be a clear demarcation between the individual instances. The boundaries between systems can be

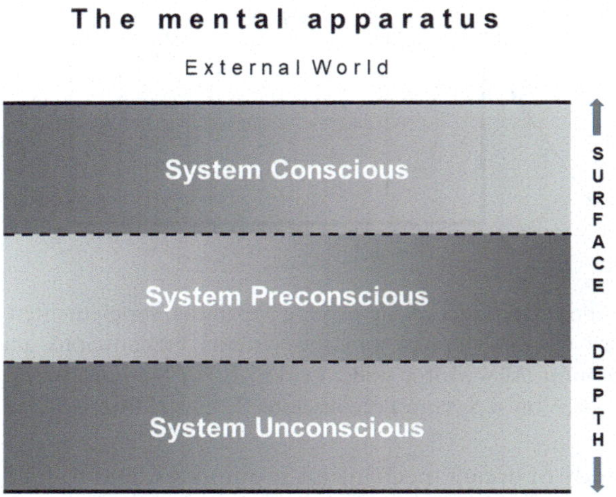

Fig. 4.4 Freud's topographical model of the psychic apparatus

imagined as a form of censorship, capable of denying or hindering access to consciousness for a desire originating from the Unconscious. This could be experienced as threatening within the individual's consciousness. This idea implies a direction of the desire, namely from the "deep" unconscious system to the surface of consciousness, as well as an inhibition of the desire by superior censors or instances in the event of a conflict. This principle has parallels with the hierarchical organisational principle of the nervous system postulated by Hughlings Jackson, according to which subordinate structures are assigned rigidly organised and automatically running functions, while superior structures are characterised by complex organisation and voluntary control (Goldstein 1995). The influence of Hughlings Jackson's theories on Freud can be traced back to Freud's work *On Aphasia* in which he postulated the principle of functional neural representation of the cortex and understood aphasia as "dis-involution", i.e. as a manifestation of an early functioning of the brain after the superior structure was destroyed by a lesion (Freud 1891; Solms and Saling 1986). Although Freud argued against a correlation between brain structure and topographic model of the psychic apparatus, there are nevertheless astonishing similarities with the Triune Brain model of evolutionary neuroethology by Paul MacLean (MacLean 1990). These correlations mainly concern functional aspects of individual cerebral associations or psychic instances.

The System Unconscious

In the topographic model of the psychic apparatus, the system Unconscious (or the Unconscious) represents that psychic instance which has no direct access to consciousness (or to the system Conscious). This topical division of the psyche in psychoanalysis into a "deep" and superior structures ultimately led to the formation of the concept of "depth psychology". The Unconscious is formed by repressed instinctual desires that are kept away from consciousness, and which—as so-called drive representations—form the psychic correlates of the drives. The process itself, which represses the contents despised by consciousness, is referred to as "repression". Although Freud assigns a somatic basis to the drive forces, they are nevertheless to be understood as a psychological construct (Freud 1905a). The drives are divided into sexual and aggressive drives based on their goal. The drive energy inherent in the sexual drive is referred to as libido, a corresponding term for the energy of the aggression drive was not conceived by Freud. The drive desires seek pleasurable satisfaction in order to minimise physical tension, in the sense of the so-called pleasure principle. This tendency of the Unconscious towards immediate satisfaction can be inhibited insofar as the passage of desires through the other psychic instances can lead to a conflict and the desires are therefore censored or transformed.

The functioning of the Unconscious according to Freud is subject to the "primary process", a form of mental processes that can be derived from the phenomena of dreams and the psychopathological symptoms (Freud 1915a). In contrast to the "secondary process", which is synonymous with conscious and rational thinking and action and is subject to the laws of logic, the primary process operates according to the principles of displacement and condensation (see chapter on the primary and secondary process).

A particular characteristic of the Unconscious is its timelessness, i.e. that processes in this system are not subject to a temporal sequence and have no reference to time. Other characteristics of the Unconscious, which can be explained primarily due to the lack of formal-logical working methods, are the absence of contradictions and the denial. The absence of contradiction means that in this system opposing elements can coexist without influencing each other, a feature known from dream images (such as the encounter with a deceased person in a dream, knowing that this person has been dead for a long time). The absence of denial in the Unconscious suggests that negation is tied to a formal thought process that is only acquired in the course of development. The predominance of infantile drive impulses or desires in the Unconscious also results in specific relationships to reality. The urge for immediate satisfaction of infantile desires necessitates a turning away from external reality ("reality principle") and functioning according to the "pleasure principle". This is characterised by the inability to tolerate denials (which are necessarily associated with the consideration of reality). In contrast to this, the "psychic reality" knows no difference between reality and fantasy, so that no distinction is made between experienced memories and fantasies. Finally, the Unconscious is characterised by the inability to abstract, i.e. that abstract concepts in the Unconscious are represented in concrete form. This phenomenon is evident in dream events, as well as a characteristic of schizophrenic thought disorders.

The System Preconscious

According to Freud's topographical model, the Preconscious, or the system Preconscious, is located between the systems Unconscious and Conscious. The boundaries of the Preconscious to the respective systems act as censorship, with the main function of the Preconscious being to protect consciousness from being overwhelmed by unconscious drive impulses.

After appropriate modification by the Preconscious, unconscious drive impulses can in principle enter consciousness. The censorship at the boundary to consciousness is much less pronounced and works less by distortion than by specific selection of contents that are allowed to enter consciousness. According to Freud's ideas, the Preconscious is to be imagined as descriptively unconscious, but it is in principle accessible to consciousness. This fact also clearly distinguishes it from the Unconscious, which is never directly accessible to consciousness. An example of a preconscious content can be a non-present memory, which can be recalled by the individual at any time. Qualitatively, the contents of the Preconscious include, on the one hand, modified drive impulses from the Unconscious, provided they have not been completely repressed, and on the other hand, the system also contains elements from experience and interaction with external reality. Ultimately, cognitive processes, such as thought processes and fantasies, take place in the Preconscious due to its close relationship with consciousness. Due to its position between the systems, the Preconscious can be seen as a kind of mediator between unconscious and conscious contents. The ability to process thoughts and access memory contents, which represent elements from real experiences, causes the

Preconscious to also function as a creative and problem-solving entity. In many cases, the performance of the Preconscious goes so far that even tasks can be accomplished that the consciousness has failed at. The close proximity of the system Preconscious to consciousness results in the former having a greater consideration of the reality principle, as well as a different mode of operation in the Preconscious, namely the secondary process. This process allows—taking into account the demands of reality—a postponement of the drive impulses striving for immediate satisfaction. The dominance of the reality principle in the Preconscious means for the individual that it can differentiate between "irreal" inner states, such as fantasies, and real events in the outside world and their sensory perception through this entity. This ability to test reality, the differentiation between the inner and outer world, is only acquired by the individual through increasing confrontation with reality during psychological development. In contrast to the primary process, the secondary process knows causality, time and logic, but no contradictions or paradoxes. The Preconscious is also crucial in the processing of psychological conflicts and involved in symptom formation.

The process that the Preconscious uses to completely suppress unwanted drive impulses is called repression. In this process, the drive impulses are again shifted back into the system Unconscious. Another possibility is the use of so-called defence mechanisms, whereby the Preconscious grants access to the drive impulses through appropriate modifications to consciousness. Symptom formation in the context of a neurotic disorder occurs when drive impulses cannot be fully pushed back into the Unconscious and no other form of drive discharge is available. In these cases, the Preconscious causes a compromise formation, which appears as a neurotic symptom. Due to its unconscious origin and the preconscious transformation, the symptom itself is experienced by the individual (i.e. by consciousness) as foreign and uncontrollable.

The System Conscious

The system Conscious, or consciousness, has a close relationship with the function of perception for Freud. He emphasises this assumption with the synonymous use of the term "perception-consciousness" (Freud 1900b). Topically, consciousness is assigned a corresponding location on the surface of the mental apparatus. The perception function refers to both external stimuli from the outside world and various internal psychological and physical stimuli. Consciousness thus encompasses from a psychoanalytic perspective the conscious experience of impressions from the environment, physical perceptions of various sensitive and sensory stimuli, but also the conscious perception of unconscious drive derivatives (which the Preconscious passed through appropriate modifications), such as in the form of certain thought processes, fantasies or daydreams. A special importance is given in this process of conscious perception to the function of attention. Only through it are internal and external stimuli consciously perceived. If attention is temporarily withdrawn from a perception, then the contents of the perception, topically speaking, are assigned to the system Preconscious. The second important function of consciousness lies in the control of voluntary movements. Naturally, in the system Conscious, the secondary process and the reality principle predominate, which, however, compared to the Preconscious, are far more comprehensive in their effect.

The Structural Model of the Mental Apparatus

The inconsistencies that arose in the course of the elaboration of the psychoanalytic theory between the clinically obtained findings and the topographical model of the mental apparatus, prompted Freud to make a corresponding modification of this metapsychological theory. It was particularly conceptual difficulties that necessitated a reordering of the mental instances. An attempt had to be made to integrate the phenomenon of fear, the aggression drive or certain psychological aspects—such as feelings of conscience or guilt—into the topographical model. However, this revision by Freud did not lead to a complete replacement of the topographical model by the so-called structural model; rather, both models, in the appropriate context, continue to be used. In a certain respect, parallels arise with the particle-wave dualism in quantum mechanics. While the central point of view, under which the topographical model was developed, represents the quality of consciousness and the ability to be conscious, the function and significance of the individual instances in the psychological conflict is in the foreground in the structural model. The psychological conflict can take place between all psychological instances, including the outside world. In the works *The Ego and the Id* (Freud 1923a) and *Inhibitions, Symptoms and Anxiety* (Freud 1926), Freud presents his second topical model—the so-called structural model—with the three psychological instances, the id, ego and superego (Fig. 4.5).

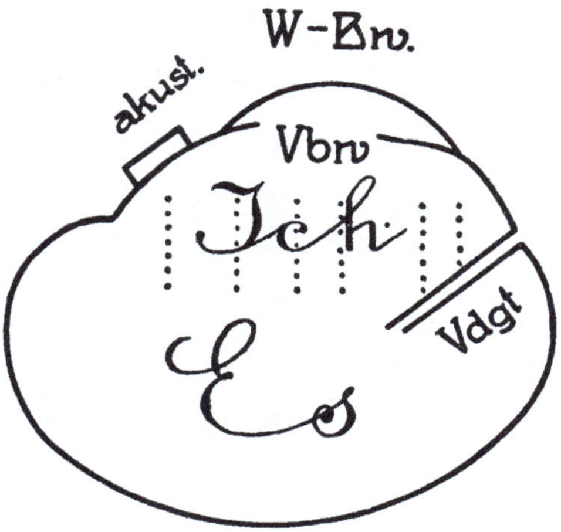

Fig. 4.5 Freud's structural model of the mental apparatus with the psychological instances "Id" (Es) and "Ego" (Ich)—(the pictorial integration of the "Superego" into the representation of the structural model only takes place in 1932, see Fig. 4.6). Vbw = preconscious, W-Bw = perception consciousness, Vdgt = repressed, akust = "hearing cap" of the "Ego" (Freud 1923a)

The ID

Freud adopts the term "id" for the description of a psychological instance from the book *The Book of the It* by the German physician Georg Groddeck. He describes the peculiarity of the It in his book as follows: "I am of the opinion that man is animated by the unknown. In him is an It, something wonderful, which regulates everything he does and what happens to him. The sentence 'I live' is only conditionally correct, it expresses a small partial phenomenon of the basic truth: 'Man is lived by the It.' ... We only know of this It what lies within our consciousness. The vast majority is inaccessible territory" (Groddeck 1923).

The instance of the id in the mental apparatus can be, in the broadest sense, compared with the system Unconscious of the topographic model. Similar to the Unconscious, the id is also hierarchically located at the lower pole, the drive pole, of the psychic apparatus or personality. Accordingly, the id contains the psychic representations of sexual and aggressive drive demands. The system Unconscious and the id do correlate in their property of being composed of unconscious and repressed elements; however, in the structural model, the entirety of the unconscious psychic is further

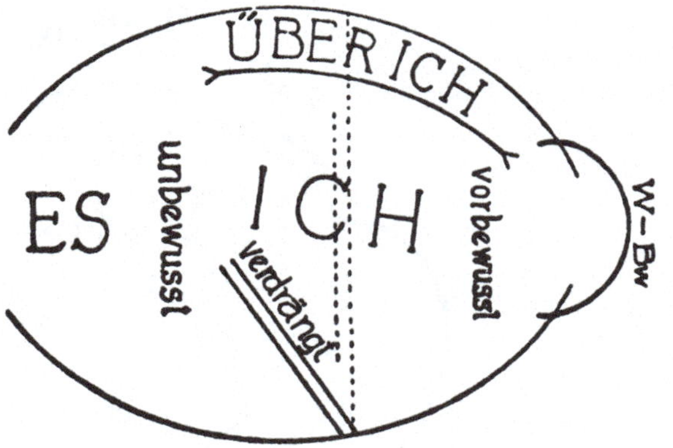

Fig. 4.6 Freud's integration of the topographical model and the structural model in a schema (Freud 1932)

defined as also including the ego. Consequently, the boundaries between id and ego are less pronounced than between the systems Unconscious and Preconscious. Finally, Freud sees the id also closer to a biological drive substrate than the topographically unconscious can be associated with the source of the drive. He clarifies this in the remark that "the id is ultimately open to the somatic" (Freud 1932). The id functions, just like the Unconscious in the topographic model, according to the pleasure principle. All processes and contents of the id are therefore also subject to the primary process. The id remains also in the structural model the central driving instance and the origin of psychic energy. Dynamically seen, this concept still results in a hierarchically organised psychic apparatus, which is driven by unconscious drive forces from the id and guided and controlled by "higher" psychic instances, the ego and the superego. This principle correlates with the central assumption of psychoanalysis that all behaviour is ultimately drive-determined (Rapaport 1970). A crucial aspect, which has been given increased consideration since the introduction of the structural model, is the developmental or (in the psychoanalytic sense) genetic point of view. As already mentioned, the boundaries between id and ego are less to be imagined as rigid censorship lines, as in the topographic model, but as flowing transitions, which blur further in the course of psychic development. This includes that all instances with increasing maturity of the individual undergo a developmental process. Hand in hand with biological maturation, the drive-related development in the id is shown as a change from oral to anal and finally genital phase.

The EGO

The introduction of the "ego" as a psychic instance within the framework of the structural model by Freud was not least due to the increasing interest of psychoanalysis in the study of defence mechanisms with the resulting lesser importance of uncovering unconscious contents. This turn made the concept of an instance necessary, which both fulfils specific defence functions and actively mediates in the psychic conflict between id demands, reality and superego contents. Compared with the topographic model, the function inherent to the Preconscious is largely attributable to the ego. Genetically considered, the ego differentiates itself in the course of psychic development from the id. Parts of the ego also remain unconscious in the descriptive sense (Freud 1923b). The Conscious is no longer a separate instance in the structural model but becomes a "sensory organ" of the ego, so that contents from the id and the superego only gain access to consciousness via the ego. The function of censorship also falls to the ego insofar as it—similar to the Preconscious—as a central instance pushes the pressing unconscious drive demands, either in modified form accessible to consciousness, or pushes dangerous contents back into the id. In addition to the defence of drive demands, the ego is also responsible for other tasks. The latter include, for example, the control of the motor skills and perception, reality testing and rational thinking, as well as the temporal order of mental processes. The shortcomings of the topographical model for a coherent metapsychological concept of anxiety were another factor for the introduction of the structural model. In this new conception, anxiety is no longer seen, as in Freud's first theory of anxiety, as a transformation of libidinous drive energy; rather, the ego uses the so-called anxiety signal, which is supposed to warn the ego of dangerous drive demands.

The anxiety signal is thus intended to protect the ego from an impending traumatic situation (Freud 1926).

The SUPEREGO

The clinical observations and phenomena of grief and depression led Freud to oppose the ego with a mental instance that simultaneously represents a function of the role model and the conscience. The term superego already implies that it is a derivative of the ego. Although prohibition and ideal functions are already hinted at in earlier works of Freud, in the so-called Ideal-Ego, the superego, with its specific tasks of prohibition and command function, is only introduced in *The Ego and the Id* as a separate instance (Freud 1923a). A pictorial integration of the superego into the structural model (Fig. 4.6) can be found in a later representation (Freud 1932).

As a derivative of the ego, the superego takes over in the course of development the commands and prohibitions originally emanating from the parents in the form of an "Internalisation", a process that is subsequently supplemented by social and cultural demands during upbringing. In the subjective experience of the individual, the superego therefore manifests primarily as a feeling of guilt but also in the form of value and moral concepts or ideals. Freud has already pointed out that the superego contains mainly "word representations", and that its contents originate from acoustic perceptions and commands. The well-known formulation, the "voice of reason", seems to illustrate this. The mentioned properties of the superego can be taken as an indication that this instance of the mental apparatus represents a late achievement of the individual, both in ontogenetic and in phylogenetic terms.

The introduction of the structural model in psychoanalysis means, as already mentioned, no abandonment of the topographical model of the mental apparatus. Both concepts prove to be valid theories in their respective context, i.e. both in the description of aspects of consciousness, as well as of intrapsychic conflicts. Both the topographical model and the structural model are based on the fundamental assumptions of psychoanalysis, i.e. the existence of the Unconscious and the differentiation between conscious and unconscious mental processes. The following schema is intended to clarify in a synoptic representation both topographical and structural aspects of the mental apparatus (Fig. 4.7).

The schematic representations of Freud's psychic apparatus remind us, not least because of his neurological background, of neural structures. Freud himself points out that the difference between the ego and the id can be compared with the contrast between reason and passion. In this view the cortex could be associated with the ego and the superego, and the id with the subcortical structures. Tarpy compared the limbic system with the id and equated the ego with the cortex (Tarpy 1977). In further integrative approaches, attempts were made to link the empirical data of evolutionary neuroethology with the psychic apparatus of psychoanalysis (Fishbein 1976).

As already described in the chapter on neuroethology, MacLean distinguishes between three different mental modes of operation, corresponding to the three macrostructures of the brain. The so-called protomentation refers to psychic functions, which are concerned with the regulation of

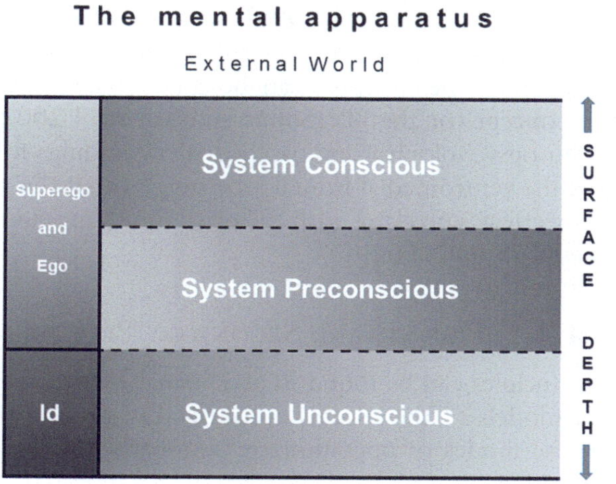

Fig. 4.7 Synoptic representation of the topographical model and the structural model of the mental apparatus according to Freud

basic behavioural routines and with certain patterns of non-verbal communication. The term "emotomentation" describes a psychic activity that regulates or influences behaviour based on the subjective perception of emotions. In analogy to these non-verbal modes of operation, the "ratiomentation" would correspond to rational, self-conscious activity where it is the only one that has the ability for linguistic communication. Without going into a topographical assignment further, parallels could be drawn here to some extent between MacLean's postulated mental modes of operation of the brain and the psychoanalytic concept of psychic instances (Ploog 2003).

The attempt to correlate the instances of the psychic apparatus and certain brain structures was first made in the context of the research field of "neuropsychoanalysis". This new scientific approach attempts to investigate insights from psychoanalysis, i.e. metapsychological concepts, using neuroscientific methods. One of the research approaches used is the so-called clinical-anatomical method, in which patients with lesions in certain brain regions are treated psychoanalytically—i.e. using free association (Kaplan-Solms and Solms 2000). Neurobiological research provides further possibilities to show correlations between brain function and psychoanalytic metapsychology. Solms and Turnbull point out, for example, that the "SEEKING system" (Panksepp 1998)—a dopaminergic neural area in the ventral tegmentum, which is particularly activated during sexual arousal and appetitive behaviours—can be functionally associated with Freud's drive concept (or the id) (Solms and Turnbull 2002a). On the other hand, neuropsychological and psychoanalytic findings from patients with lesions in the ventromedial frontal lobe suggest that this region, due to its inhibitory effect, correlates with the drive-inhibiting function of the ego (Solms and Turnbull 2002b).

4.2 Primary and Secondary Process

Hierarchical principles can be found in psychoanalytic theory not only in the structural models of the psychic apparatus but also in its modes of operation. These modes of operation are summarised in psychoanalysis under the term of economy.

The economic perspective of psychoanalytic theory is based on the fundamental assumption that all behaviour discharges psychic energy and that behaviour is regulated by psychic energy (Rapaport 1970).

The two modes of operation of the psychic apparatus are referred to by Freud as primary process and secondary process. The two terms already indicate a hierarchical organisation through their naming. The primary process is therefore that mode of operation that is already present from the beginning of the psychic development of the individual. The secondary process, on the other hand, operates according to principles that only become effective in the individual during the course of development and in the course of dealing with reality. The concept of primary and secondary process was already formulated by Freud in 1900 in the interpretation of dreams and remained an unchanged central reference point in metapsychology. The study of dream mechanisms was finally what prompted Freud to assume a different form of mental functioning than that of rational thought processes. This other mental mode of operation, the primary process, is characterised by two principal mechanisms that can be found not only in dreams but also in the symptom formation of neurotic disorders: displacement and condensation.

In displacement, the meaning or intensity of an idea is subtracted from it and transferred to another, originally less intense idea, which is associated with the first. This results in a transformation of psychic value from an emotionally significant idea to a more insignificant content. In the symptom formation in the context of phobia, e.g. agoraphobia, the mechanism of displacement is involved insofar as here for the person threatening, unconscious drive impulses are "displaced" onto an insignificant but associatively connected content (or object).

The associative connection means that the displacement, following the principles of the unconscious, is subject neither to logical laws nor to formal rules.

The mechanism of condensation was first introduced by Freud in *The Interpretation of Dreams.* He describes the dream phenomenon whereby various elements or idea contents can be combined into a unit. Examples of this are the well-known "composite persons" in dreams, which represent a kind of conglomerate of various individuals, and thus illustrate a common element of these people in a "condensed" form. Freud already pointed out that in condensation there is always a fusion of those images or thought contents that have similarities and are therefore not randomly selected. This fact can be used as an argument against the repeated accusation that dreams are only a random sequence of images. In Freud's view, the mechanism of condensation rather "represents a sophisticated technique that cleverly exploits random similarities between two objects". A vivid idea of the workings of the condensation process is provided by a special form of computer graphics. In these graphics, content-related images are arranged according to a given algorithm in such a way that from the mosaic a new (but content-related to the individual images) overall image emerges (Photomosaic©). This process, or this image, resembles in many respects dream images or dream contents, which are condensed from many individual images (or associatively related thought contents).

Apart from its effectiveness in dream events, condensation is also one of the essential elements in humour or in mistakes, such as in a slip of the tongue.

Economically or energetically considered, the primary process is characterised by a great mobility of the so-called cathectic intensities. Only through this is it possible in the displacement that one idea can give all of its cathexis to another. Similarly, in condensation, the entire cathexis of several ideas is combined into one. The primary process strives to discharge the excitations, in order to establish a "perception identity" with the thus collected excitation size, while in the secondary process an attempt is made to achieve a "thought identity" through discharge inhibition. Freud sees the dream as an image of the primary process, since through the regressive character of the dream image a perception identity is established. In his work *Formulations on the Two Principles of Mental Functioning*, he draws parallels between primary process and pleasure principle, as well as between secondary process and reality principle (Freud 1911a).

The formal characteristics of the primary process, which include timelessness and the concretistic mode of operation as well as the absence of contradictions and negations, have already been discussed in the explanations about the Unconscious. Topographically considered, primary process and secondary process can be clearly assigned to certain structures of the psychic apparatus. Thus, the primary process characterises the system Unconscious and the secondary process the system Preconscious-Conscious.

The Empirical Dimension of the Primary Process

The first experiments to investigate the primary process using empirical methods were undertaken by Otto Pötzl (1917). He presented subliminal visual representations to subjects using a tachistoscope, a visual projection device that produces images below the threshold of perception. The test subjects were then asked to describe their visual impressions and the subsequent dreams or to record the remembered dream images. The results showed that the remembered dream images of the unconsciously registered stimuli were often subjected to that transformation and distortion which Freud referred to as condensation, displacement and symbolic representation, and thus can be considered as manifestations of the primary process. Fisher replicated and extended this experimental technique by offering both subliminal and supraliminal stimuli and was able to show that exclusively the subliminal stimuli were subjected to a primary process-like transformation (Fisher 1960). Studies on split-brain patients, i.e. individuals whose hemispheres were separated by callosotomy, also influenced subsequent psychoanalytic research. Galin was the first to point out the parallels between the psychoanalytic theory of primary and secondary process and the specialisation of the cerebral hemispheres (Galin 1974). The assumption that the left hemisphere with logical and real properties, i.e. the secondary process, and the right hemisphere with emotional and symbolising elements, i.e. the primary process, are related, was later often referred to as reductionistic, and accordingly modified or partially revised. Contrary to this view is the argument that both hemispheres are connected by the corpus callosum, thus ensuring a continuous and simultaneous perception and organisation of experiences in the various cognitive modalities. Accordingly, at least in the adult individual, a dynamic coexistence of primary process and secondary process can be assumed (McLaughlin 1978).

Using a modified form of Pötzl's subliminal stimulation technique, Shevrin conducted a series of investigations into the functioning of the primary process (Shevrin and Luborsky 1960). Among other things, he shortened the stimulation duration of the images to 1 millisecond to definitely exclude conscious visual perception. Using a special form of stimulation, he was able to induce both primary process- and secondary process-related effects. The stimulus itself consisted of a pictorial representation of a rebus (in reference to Freud's comparison of the dream's picture language with a rebus), namely the image of a writing pen (pen) on a knee (knee), which together formed the word "penny". From the associations induced after the stimulations, several results were obtained: (1) Secondary process-related effects, characterised by categorical associations (e.g. for pen "ink" or "paper" and for knee "foot" or "leg"), (2) primary process-related effects, which were characterised by sound associations to the individual words (e.g. "open" for pen) and (3) primary process-related condensations of the words pen and knee to "penny", which led to an effective new word formation.

In another series of experiments, the effect of subliminal stimulation on the various stages of sleep was investigated (Shevrin and Fisher 1967). The results showed primary process-related effects in associations immediately after a REM phase, while secondary process-related effects occurred in associations after non-REM phases. These constellations would be indirect confirmations for the psychoanalytic theory that in dreams (i.e. during REM phases) the functioning of the primary process is predominant.

In view of the scientific-theoretical demand that the proof of a theory must be achieved by methods that are independent of the theory itself, other psychoanalysts also attempted to provide experimental evidence of the existence of the primary process through empirical studies. In this context, using the subliminal stimulation method, the ability to categorise pictorial material—both in the subliminal and supraliminal mode—was investigated (Brakel et al. 2000). Basic preparatory work on the topic of categorisation and similarities was carried out by Smith and Medin (Smith 1981). The authors distinguish between "attributive" and "relational" similarity. An "attributive" similarity refers to concrete, specific and superficial characteristics, while "relational" describes a similarity of relationships among attributes (Brakel et al. 2000). An example will further clarify the terms: If the master figure consists of an image on which a disc is located between two triangles, then a relationally similar image would be, for example, a star between two squares. On the other hand, an image in which, for example, two triangles are on top of each other and a disc is somewhere next to them would be an attributionally similar image. When subjects are given the choice of selecting one of two images in relation to the similarity to a given image (master figure), the relationally similar image is more often assigned to the master figure (Medin 1990).

This experimental paradigm was expanded by Brakel to include additional subliminal stimulation, in order to investigate whether the decisions regarding the similarity features are subject to primary or secondary process-related aspects (Brakel et al. 2000). The authors of the study assumed that similarity decisions based on relational criteria are assigned to the secondary process, and decisions based on attributional basis are assigned to the primary process. The study showed that the subjects, despite the complexity of the subliminal stimuli, were able to grasp the presented image and make decisions regarding the similarity of the stimuli. The experiments also demonstrated that images with attributional similarity (thus assignable to the primary process) were chosen significantly more often in similarity decisions with subliminal stimulation. Although the data collected were not obtained using psychoanalytic methods, they nevertheless deal with formal aspects of primary process-related thinking, particularly condensation and displacement and the associated mental manifestations. From this perspective, results from controlled studies provide independent evidence supporting basic concepts of metapsychology or psychoanalysis.

The previously mentioned categorisation test based on similarity criteria offers further possibilities to investigate or confirm characteristic aspects of the primary process. To test the primary process-determined thinking organisation of preschool children, they were asked in a standardised experimental setup to select the image that they believed was most similar to the "master figure". It was found that the decisions—correlating with psychoanalytic theory—were made predominantly based on attributional aspects, which is consistent with the predominance of the primary process (Brakel 2004). A similar, significant shift towards attributional categorisation was also found by Brakel in patients with previously objectively lowered anxiety levels. These results are also compatible with the psychoanalytic observation that patients with a predisposition to anxiety states increasingly exhibit functions that correspond to the primary process (Brakel 2004).

The question of the function of the primary process was answered by Freud in that he considered it an early form of thinking that, in the course of phylogeny, now manifests itself mainly in dreams due to a lack of function. Meanwhile, numerous empirical studies provide evidence that unconscious information processing actually exists and that unconscious contents can influence conscious behaviour. Regarding subliminal perception, clear neurophysiological data and concepts are available (Dixon 1971; Dehaene and Changeux 2004). The assumption of the existence of subliminal perceptions is supported by the fact that afferent pathways of the lemniscal system (pathways that lead sensory information to the cortex) transmit information faster than those of the extralemniscal system (which also includes the ascending reticular system necessary for consciousness). For this reason, it is possible for an incoming stimulus to reach the cortex (so that the cortex simultaneously inhibits the reticular system), even before a neural impulse reaches the reticular system, and thus would have induced conscious perception. This model of unconscious information processing, designed for external stimuli, could in principle also be envisaged for internal stimuli (motivations, desires, needs). Such a mechanism could provide the neurobiological basis for an explanatory model of unconscious conflict situations (Shevrin and Dickman 2003).

The evidence of a neural substrate for unconscious information processing subsequently raises the question of the biological function. In relation to the primary process, the question arises of the evolutionary significance of a mode of operation of the psychic apparatus, which among other things employs the mechanisms of condensation, displacement or symbolic representation. Following Freud's assumptions on this question, comparative behavioural research could provide important contributions. Ethology owes the discovery of the significance of so-called key stimuli (Franck 1997). These represent visual or acoustic stimuli that are simply structured (i.e. consist of few elements) and at the same time are striking and unambiguous. Such stimuli have the biological function in the animal kingdom of triggering instinctual actions. In addition, the stimuli have symbolic character. In dummy experiments, Tinbergen was able to show that instinctual actions in animals could also be triggered by extreme abstractions of the key stimulus, and that through extreme schematisations or symbolisations (e.g. replacement of the mother animal's beak with a rod with three points) even more actions could be induced (Tinbergen

1951). If one were to associate such schematisations or symbolisations with the mechanisms of displacement and condensation, this form of perception organisation could also be compared in the broadest sense with the modes of operation of the primary process. According to this assumption, perception, which takes place according to the laws of the primary process, would have the biological function of activating the motivation system and thus inducing specific behavioural patterns, with the ability to symbolise being seen as an extension or safeguarding of the stimulus effect. Similar key stimuli with subsequent instinctual action in the form of smiling can be detected in infants aged three to six months, i.e. at an age when, according to psychoanalytic view, the primary process predominates (Spitz 1965). Even in the infant, the reaction (the smile) can be intensified by extreme schematisation of the human face. These phenomena are no longer detectable with increasing maturation of the toddler.

The previous explanations, which would characterise the primary process as a phylogenetic atavism, would be compatible with Paul MacLean's phylogenetic-evolutionary concept of brain function. MacLean himself compared the non-linguistic, so-called prosematic, form of communication of the old brain structures, i.e. the reptilian brain and the limbic system, with the functional principles of Freud's primary process (MacLean 1990). The thought organisation of the primary process and its mechanisms (condensation, displacement, symbolisation) would, according to these ideas, be the "language" of the older brain formations.

Functions specific to the primary process would manifest themselves when these systems are activated in their specific "language", e.g. as symbolic representation of dream contents, or in the form of neurotic symptoms. The simultaneous functioning of the brain on different evolutionary levels would correlate with a constant coexistence and mutual influence of primary and secondary processes and explain manifestations of the primary process in the form of mistakes, jokes, neurotic symptoms, dreams or art as an integral part of human experience.

The functioning of the secondary process arises, from an evolutionary perspective, due to the increasing confrontation of the individual with the realities of life. Accordingly, the secondary process is characterised by properties that oppose the elements of the primary process. Thus, the laws of logic and causality, as well as the sense of time, or the incompatibility with contradictions and paradoxes are inherent to the secondary process. Upon reaching the "thinking identity" in the secondary process, reality-appropriate mental performances, such as rational thinking, attention and voluntary actions, are enabled. According to Freud, these abilities presuppose a tendency towards the "binding of psychic energy", i.e. that in rational thinking only small amounts of psychic energy are shifted, quite contrary to the primary process, in which the psychic energy is freely movable and easily shiftable. A significant feature of the secondary process is the ability to use language. With it, words are no longer processed concretely but as a tool for the symbolic representation of abstract ideas and concepts.

4.3 The Concept of Regression

Regression, in the general sense, means a return to a previous state. In psychoanalysis, it refers to a return to an earlier stage of development or an earlier stage of psychic function. The concept of regression is thus implicitly associated with a hierarchical organisation of the psyche, or psychic development, in which there is a progression from a primitive to a complex stage. Regression itself recalls the principle of dissolution postulated by Hughlings Jackson and Herbert Spencer, i.e. the reversal of an evolutionary process with the loss of the superordinate layers and the associated exposure of subordinate structures (Goldstein 1995).

The concept of regression was already used by Freud in his neurological work *On Aphasia* (1891). Following Hughlings Jackson's principle of dissolution, Freud saw in the symptoms of aphasia a state that already existed or was passed through at the beginning of language development (Freud 1891). In the *Interpretation of Dreams* (1900a), Freud introduces the concept of hallucinatory wish fulfilment, which represents a primitive form of mental activity and is abandoned by the infant very early in development. Only in dreams and in psychosis does a return to this lower psychic functioning occur, in the sense of a regression. This form of regression presupposes a hierarchical organisation of the psychic apparatus, as realised in the topographical model. For this reason, in the case of dreaming, hallucinations or psychosis, one also speaks of a "topographical or topical regression", i.e. a return from the system Vbw-Bw to the system Ubw, or from the secondary process to the primary process.

In his work *Three Essays on the Theory of Sexuality* (Freud 1905b), Freud refers to another form of regression, which is to be considered in a genetic context. He speaks of a "return" of the libido to love objects of childhood or to earlier stages of libido development, i.e. a regression that primarily does not occur within the psychic apparatus but which occurs in the course of mental development. This so-called temporal regression is to be understood as the return of the subject to stages that have already been passed through in development. Examples could be hysteria and obsessive-compulsive neurosis. In the former, from a psychoanalytic perspective, there is a regression of the libido to the primary sexual objects of childhood, the obsessive-compulsive regresses even to a pre-genital stage of libido development, namely to the anal stage. The temporal regression can refer to stages of libido development, object relations or other aspects of development.

Finally, the term "formal regression" describes a return to more primitive forms of expression and behaviour. A formal regression could therefore be spoken of in the transition from secondary to primary process, or from thought identity to perceptual identity. For Freud, the return to earlier stages of mental function was a fundamental characteristic of mental illnesses. Regardless of the different forms, regression represents a core concept of psychoanalytic theory with regard to the genesis of neuroses and symptom formation (Jackson 1969).

Gedo and Goldberg, following Freud, designed a hierarchical model of psychological development, which describes the degree of maturity of the mental apparatus depending on the phase-like temporal development of the individual (Gedo and Goldberg 1973). Since this model also takes into account aspects of relationships to objects (i.e. the ability to distinguish between self and objects), it allows for a more precise formalised description of the regression stage of an individual. This concept therefore also includes further developments of psychoanalytic theory (narcissism theory, object relations theory, self-psychology) which provide a more detailed picture of mental functioning between reflex arc model and structural model (Gedo and Goldberg 1973).

The horizontal axis of the scheme is divided into phases, which can include various development-related aspects.

Phase I extends from the time of birth to the acquisition of the ability to cognitively differentiate between self and object.

Phase II extends from the last mentioned point to the functional separation of the self from the object.

Phase III extends from the time of consolidation of a coherent self to the formation of the superego.

Phase IV extends from the formation of the superego to the differentiation of the ego.

Phase V represents the point in time of the fully differentiated mental apparatus.

The temporal sequence of the individual phases schematically reflects the normal development of personality. Mental conflicts would within this development scheme be imposed by the resulting regression as autonomous areas that do not participate in the normal personality development. In this hierarchical development concept, the models of the mental apparatus are aligned on the vertical axis and assigned to the phase-like development process on the horizontal axis. According to Gedo and Goldberg, this development-related hierarchical model should not be understood as a rigid system. For example, the representation of developmental phases can be modified by further subdivision into subgroups to obtain a more accurate picture of mental functioning.

The Phylogenetic Regression

Some authors hold the view that in addition to the previously discussed ontogenetic regression, a phylogenetic regression also exists (Meerlo 1962; Bailey 1978). Merlo sees regression as a process that, as a result of danger, fear or stress, is always both an ontogenetic and phylogenetic step backwards. Regression leads to acquired behaviours and habits losing importance compared to older ways of functioning. Another characteristic of regression is that it occurs more easily than its opposite, progression (Meerlo 1962). In addition to the general characteristics of a phylogenetic regression, which include a decrease in the importance of cultural and moral achievements of humans, the question arises of specific behaviour-related characteristics of phylogenetic regression.

Some of the properties to be mentioned in this regard would be the reduction of inhibitory factors associated with phylogenetic regression, as well as an increase in aggression and the pursuit of pleasure (Bailey 1978). The loss of inhibitions, which can be observed daily in the context of violent confrontations, is also associated with a change in response behaviour—in the sense of the "all-or-nothing" principle. This means that behaviour becomes more automatic and is less subject to the secondary process-related aspects of the ego and superego. In this sense, disinhibition is equivalent to behavioural atavism. Danger seems to have played a significant role in phylogenesis in the development of regressive behaviours; moreover, danger seems to lead to more extreme forms of regression (Schur 1960). The human capacity for aggression, up to the most extreme forms of violence—as a phenomenon of phylogenetic regression—does not need to be further pointed out in view of its ubiquity within society. Ontogenetically, according to Freud, the pleasure principle is the primary mode of operation of the psychic apparatus, which is only replaced by the reality principle with increasing development. This concept already implicitly contains a genetic-hierarchical organisation. Freud's idea correlates with the assumption that in phylogenetic regression, the pursuit of pleasure comes to the fore again.

Rado sees hedonistic or pleasure-seeking behaviour as a phenomenon of the phylogenetically most primitive part of the psychic system (Rado 1969).

Another characteristic of phylogenetic regression is the importance of non-verbal communication. This is particularly evident in humans in connection with aggressive, sexual or danger-associated behaviours. Although the information content of non-verbal expressions is low compared to language, they seem to have great communicative importance. In a certain sense, non-verbal vocalisations can be seen as regressive manifestations of instinctive expressions (Bailey 1978). A neurobiological and ethological parallel arises in this context insofar as, according to MacLean's Triune Brain theory, the archaic macrostructures of the brain (R-complex and paleo-mammalian brain) are only capable of non-verbal communication. Increased tendencies towards cooperation and a willingness to recognise authorities within the group are also characteristics of phylogenetic regression (Bailey 1978).

4.4 Phylogeny and Ontogeny in Psychoanalysis

The importance that Freud attributed in psychoanalysis to phylogenetic and ontogenetic aspects—and their connections—can be traced in many parts of his works. Phylogenetic and ontogenetic development is always understood as a stage-like maturation process that leads to a hierarchically organised functional structure. This does not mean that earlier stages are replaced by more mature developmental stages, but that these overlay the earlier stages. This hierarchical structure also explains the return from a superior to a subordinate stage, in the sense of a regression.

Freud's interest in the connections between phylogeny and ontogeny is not least due to the then prevailing spirit of biology. Freud was not only influenced by Darwin (Ritvo 1993) but especially by Haeckel (Sulloway

1979) and Lamarck (Ritvo 1965). Long before the discovery of the psychosexual developmental stages—the oral, anal and genital phase—Freud suspected a phylogenetic cause in the mechanism of repression. He illustrates his theory using the change in the significance of smell in onto- and phylogenesis. Accordingly, the development of upright walking in humans was associated with a functional change in the olfactory organ, which was no longer facing the ground. According to Freud, this change also led to olfactory sensations, which were previously interesting for the individual, now being perceived as repulsive and disgusting. In analogy to this phylogenetic process, there would be a repression of originally pleasurable olfactory tendencies during the course of childhood sexual development. In this respect, there would be a repetition of phylogenetic aspects in ontogeny (Freud 1897). This recapitulation of phylogeny in ontogeny essentially corresponds to Haeckel's biogenetic principle (Haeckel 1874), although Freud's assumptions concern psychic aspects. At the same time, this view contains the concept of an inheritance of acquired mental characteristics in the sense of Lamarck's theory of evolution.

In the *Three Essays on the Theory of Sexuality*, Freud refers to the animal origins of the oral and anal stages of psychosexual development, comparing them to "relapses into early animal states" (Freud 1905c). The biological roots of the latter stages, according to these views, still date back to the times of quadrupedal locomotion, when taste and smell predominated, and sight was not yet a significant sensory function. Haeckel's biogenetic principle refers to physical characteristics that are repeated in the ontogeny of the individual. An important aspect, however, is that this physical repetition of phylogeny in ontogeny is only a transient phenomenon. The archaic morphological structures, which the embryos of many species share at the beginning of development, are replaced in the course of maturation by more differentiated morphological structures. This means that the physical recapitulation in ontogenesis is only a progression through earlier stages of development. On the adult human body, there are only barely recognisable residues of the phylogenetic heritage, such as in the form of the so-called Darwin's tubercle on the ear, a protrusion of the ear rim, which represents a rudiment of the animal ear tip. The situation is different with the mental recapitulation of phylogeny in the development of the individual postulated by Freud, as in this case the individual stages can exist side by side. Although there is also a chronological progression through the individual stages, the older stages are not replaced by the subsequent ones, but rather, there is an inhibition of the older stages by the subsequent ones. In a certain sense, parallels arise here with the so-called primitive reflexes. They too represent the phylogenetic heritage of archaic, life-necessary reflexes, which are present in ontogenesis until late infancy and are inhibited in the course of further maturation by higher cerebral structures. The primitive reflexes are also not replaced but continue to exist in a latent (inhibited) form and reappear through lesions of higher cerebral structures.

From a psychoanalytic perspective, neurosis involves a regression to earlier stages of psychosexual development. The neurotic is not only fixated in his psychosexual maturation due to the regression to a childlike stage of sexuality, but he also represents a state of phylogenetic regression, i.e. he shows characteristics in his behaviour that were once characteristic of human ancestors. In *Psycho-analytic Notes on an Autobiographical Account of a Case of Paranoia*, Freud points to the parallels between the mental life of children, contemporary primitive peoples and neurotics. They all recapitulate archaic mental states of human ancestors and share, in his view, a common phylogenetic heritage (Freud 1911b).

In the treatise *Overview of the Transference Neuroses*, Freud goes a step further and tries to establish a correlation between the characteristics of certain psychopathological states and various phylogenetic developmental stages (Freud 1915b). Although he himself describes these statements as "phylogenetic fantasy", they represent an attempt to establish connections between psychopathological entities, as well as ontogenesis and phylogenesis. Freud creates a chronology of neurotic and psychotic (narcissistic) disorders in ontogenesis, corresponding to the extent of their regression to early psychosexual stages. Anxiety hysteria thus represents the earliest neurosis, followed by conversion hysteria, which is observed from about the fourth year of life. Obsessive-compulsive disorders only occur somewhat later, in prepuberty. The psychotic (originally referred to by Freud as narcissistic) disorders are not described in childhood. These include schizophrenia (Dementia praecox) as a classic disease of puberty, and paranoia or manic-depressive disorders (Melancholia-Mania), which only become manifest in the years of maturity. This chronological listing implies that the regression in the respective disease reaches further back, the later the neurosis or psychosis occurs in the development of the individual.

Based on the assignment of mental disorder and in the psychosexual stage of ontogenesis, Freud also suspects a correlation between phylogenetic stages of human development and individual neurotic diseases. Phenomena of anxiety hysteria or obsessive-compulsive disorder represent, according to these views, mental states that were common to all people in the Ice Age, while in the present they are only characteristics of neurotic disorders.

Freud goes into more detail on the similarities of the mental life of children, primitive peoples and neurotics in *Totem and Taboo* (Freud 1913a). The way of thinking of children, characterised by the so-called "omnipotence of thoughts", shows striking parallels with the so-called animism of primitive tribes. Both forms of thought are characterised by the idea that thoughts are capable of changing the environment. At the same time, inanimate objects of the outside world or animals are personified. It also turns out that the pathological behaviour of neurotics consistently corresponds to the behavioural norms of primitive peoples.

The example of the incest taboo makes it clear that the growing child must, in the course of normal maturation, detach itself from the originally incestuous love objects. In the case of neurotic development, this detachment fails, resulting in a psychological infantilism and an unconscious incestuous attachment to the love objects of childhood persists. A parallel to the behaviour of primitive peoples arises insofar as they also cling to their incestuous desires. To protect against these strong desires, these peoples erect taboos to make real incest impossible. What persists in the neurotic in an unconscious form and ultimately determines behaviour, is experienced by primitive peoples as a real threat that must be countered with the strictest measures. Both cases can, in this view, be considered a kind of mental atavism.

In the case of the existence of a mental atavism in ontogenesis, a corresponding correlate would also be demanded for one of the most important phases of psychosexual development, the Oedipus complex (the conflict situation of the child with the same-sex parent, due to his incestuous desires towards the opposite-sex parent) in the phylogeny of man. The phenomena of taboo and totemism, i.e. behaviours found in primitive peoples—and thus also assumed in the behavioural repertoire of human ancestors—form the starting point for an explanatory model for Freud in this respect. Based on ethological observations regarding the hierarchical social structure of primates, the cohabitation of human ancestors could be imagined in the form of a patriarchal horde. This horde would, analogous to the primates, be dominated by a male individual, who also had the sexual prerogative over the female members of the horde. According to Freud, the banished sons of the horde father would have subsequently joined forces, killed and eaten the father. The patricide would have induced such feelings of guilt in the sons that they avoided sexual contact with the female members of the clan, and identified the dead father with an animal (the totem), to which sacrifices were then made, and which was not to be harmed. If the totem animal actually corresponded to the father, the two resulting commandments (taboos), namely not to kill the totem animal and to avoid incestuous contact with female members of the horde, would exactly match the content of the story of Oedipus, in which the son killed his father and married his mother. The same conflict from the phylogeny of humanity is also repeated in this way in the ontogeny, in the form of the Oedipus complex (Freud 1913b). These explanations also contain the idea that in the mental life of children and neurotics, mental conflicts are recapitulated, which once corresponded to a real conflict.

The repetition of phylogeny in ontogeny can in principle be extended to all cognitive abilities and developments of humans. Accordingly, the entire cultural development of humanity could also be traced in the maturation of the individual. In this view, the emergence of cultures, which from the perspective of psychoanalysis only become possible through renunciation of drives, is reflected in the maturation of the individual. This also only meets societal demands through the taming of sexual and aggressive drives (Freud 1930).

The assumption of a recapitulation of archaic mental and behaviour-related phenomena in ontogeny not only correlates—at least in principle—with Haeckel's theory (who limited recapitulation only to morphological features) but also includes an inheritance of acquired characteristics. This view, coined by Lamarck and considered outdated after the emergence of Darwin's theory of evolution, has only regained importance in recent years through the new scientific discipline of epigenetics. Epigenetics deals with inheritable changes in phenotype through mechanisms that do not involve a change in DNA. Epigenetic changes have also been observed under the influence of environmental factors, so it is suspected that this mechanism may possibly be involved in the short-term adaptation of a species to changed environmental conditions (Bird 2007). This supports Lamarck's theory that a transmission of acquired characteristics between generations can also take place. Since epigenetic changes in animal models have now been demonstrated even in relation to fear, depression and addiction (Jakobsson et al. 2008), Freud's Neo-Lamarckian theories also appear in a new light.

Bibliography

Bailey KG (1978) The concept of phylogenetic regression. J. Amer. Acad. Psychoanal. 6: 5–35.

Bird A (2007) Perceptions of epigenetics. Nature 447: 396–398.

Brakel LAW (2004) The psychoanalytic assumption of the primary process: extra-psychoanalytic evidence and findings. J. American Psychoanal. Ass. 52: 1131–1161.

Brakel LAW, Kleinsorge S, Snodgrass M, Shevrin H (2000) The primary process and the unconscious: experimental evidence supporting two psychoanalytic presuppositions. Int. J. Psychoanal. 81: 553–569.

Dehaene S, Changeux JP (2004) Neural mechanisms for access to consciousness. In: Gazzaniga MS (Ed) The Cognitive Neurosciences III. The MIT Press, Cambridge, London. pp 1145–1157.

Dixon F (1971) Subliminal perception: The nature of a controversy. McGraw-Hill, London.

Fishbein HD (1976) Evolution, development, and children's learning. Goodyear, Pacific Palisades, California.

Fisher C (1960) Subliminal and supraliminal influences on dreams. American J. Psychiatr. 116: 1009–1017.

Franck D (1997) Verhaltensbiologie. 3. Auflage, Thieme Verlag, Stuttgart, New York, pp 11–14.

Freud S (1891) Zur Auffassung der Aphasien. Eine kritische Studie. Fischer Verlag, Frankfurt am Main 2001.

Freud S (1897) Brief 146. In: Masson JM (Ed) Sigmund Freud – Briefe an Wilhelm Fließ. S. Fischer. Frankfurt am Main, 1986, pp 301–305.

Freud S (1900a) Die Traumdeutung. Gesammelte Werke, II/III, Fischer Verlag, Frankfurt am Main 1999, pp 542–546.

Freud S (1900b) Die Traumdeutung. Gesammelte Werke, II/III, Fischer Verlag, Frankfurt am Main 1999, pp 620–622.

Freud S (1905a) Drei Abhandlungen zur Sexualtheorie. Gesammelte Werke, V, Fischer Verlag, Frankfurt am Main 1999, p 67.

Freud S (1905b) Drei Abhandlungen zur Sexualtheorie. Gesammelte Werke, V, Fischer Verlag, Frankfurt am Main 1999, pp 69–70.

Freud S (1905c) Drei Abhandlungen zur Sexualtheorie. Gesammelte Werke, V, Fischer Verlag, Frankfurt am Main 1999, p 98.

Freud S (1911a) Formulierungen über die zwei Prinzipien des psychischen Geschehens. Gesammelte Werke, VIII, Fischer Verlag, Frankfurt am Main 1999, pp 231–232.

Freud S (1911b) Psychoanalytische Bemerkungen über einen autobiographisch beschriebenen Fall von Paranoia, VIII, Fischer Verlag, Frankfurt am Main 1999, pp 319–320.

Freud S (1913a) Totem und Tabu. Gesammelte Werke, IX, Fischer Verlag, Frankfurt am Main 1999.

Freud S (1913b) Totem und Tabu. Gesammelte Werke, IX, Fischer Verlag, Frankfurt am Main 1999, pp 122–194.
Freud S (1915a) Das Unbewusste. Gesammelte Werke, X, Fischer Verlag, Frankfurt am Main 1999, pp 264–303.
Freud S (1915b) Übersicht der Übertragungsneurosen. Grubrich-Simitis I (Ed). S. Fischer Verlag, Frankfurt am Main 1985, pp 70–81.
Freud S (1923a) Das Ich und das Es. Gesammelte Werke, XIII, Fischer Verlag, Frankfurt am Main 1999, pp 235–290.
Freud S (1923b) Das Ich und das Es. Gesammelte Werke, XIII, Fischer Verlag, Frankfurt am Main 1999, p 244.
Freud S (1926) Hemmung, Symptom und Angst. Gesammelte Werke, XIV, Fischer Verlag, Frankfurt am Main 1999, pp 111–206.
Freud S (1930) Das Unbehagen in der Kultur. Gesammelte Werke, XIV, Fischer Verlag, Frankfurt am Main 1999, pp 421–506.
Freud S (1932) Neue Folge der Vorlesungen zur Einführung in die Psychoanalyse. Gesammelte Werke, XV, Fischer Verlag, Frankfurt am Main 1999, p 80.
Galin D (1974) Implications for Psychiatry of left and right cerebral specialisation: A neurophysiological context for unconscious processes. Arch. Gen. Psychiat. 31: 572–583.
Gedo JE, Goldberg A (1973) Models of the mind. A psychoanalytic theory. The University of Chicago Press, Chicago, London pp 73–150.
Goldstein RG (1995) The higher and lower in mental life: an essay on J. Hughlings Jackson and Freud. J. Amer. Psychoanal. Assn. 43: 495–515.
Groddeck G (1923) Das Buch vom Es. Fischer Verlag, Frankfurt am Main 1979, p 7.
Haeckel E (1874) Anthropogenie oder Entwicklungsgeschichte des Menschen. „Engelmann", Leipzig.
Jackson SW (1969) The history of Freud's concepts of regression. J. American Psycho- anal. Assoc. 17: 743–784.
Jakobsson J, Cordero MI, Bisaz R, Groner AC, et al. (2008) KAP1-mediated epigenetic repression in the forebrain modulates behavioural vulnerability to stress. Neuron 60: 818–831.
Kandel ER (2006) Psychiatrie, Psychoanalyse und die neue Biologie des Geistes. Suhrkamp, Frankfurt am Main, pp 119–183.
Kaplan-Solms K, Solms M (2000) Clinical studies in Neuro-Psychoanalysis. Karnac Books, London.
Loch W (1999) Die Krankheitslehre der Psychoanalyse. S. Hirzel Verlag, Stuttgart, Leipzig, pp 25–30.
MacLean PD (1975) On the evolution of three mentalities. Man-Environment-Systems 5: 213–224.
MacLean PD (1990) The Triune Brain in evolution. Role in paleocerebral functions. Plenum Press, New-York, London.

McLaughlin JT (1978) Primary and secondary process in the context of cerebral hemispheric specialisation. Psychoanalytic Q. 47: 237–266.

Medin D (1990) Similarity involving attributes and relations: judgements of similarity and difference are not inverse. Psychol. Sci. 1: 64–69.

Meerlo JAM (1962) The dual meaning of human regression. Psychoanal. Rev. 49: 77–86.

Panksepp J (1998) Affective Neuroscience: The foundations of human and animal emotions. Oxford University Press, New York and Oxford.

Ploog DW (2003) The place of the triune brain in psychiatry. Physiology & Behavior 79: 487–493.

Pötzl O (1917) Experimentell erregte Traumbilder und ihre Beziehungen zum indirekten Sehen. Zeitschr. Ges. Neurol. Psychiatr. 37: 278–349.

Rado S (1969) Adaptational psychodynamics: motivation and control. Science House, New York.

Rapaport D (1970) Die Struktur der psychoanalytischen Theorie. Ernst Klett Verlag, Stuttgart, pp 43–77.

Ritvo LB (1965) Darwin as the source of Freud's Neo-Lamarckianism. J. Amer. Psychoanal. Assn. 18: 195–208.

Ritvo LB (1993) Darwin's influence on Freud: A tale of two sciences. Yale Univ. Press, New Haven & London.

Schur M (1960) Phylogenesis and ontogenesis of affect- and structure-formation and the phenomenon of repetition compulsion. Int. J. Psychoanal. 41: 275–287.

Shevrin H, Luborsky L (1960) The rebus technique: a method for studying primary process transformations of briefly exposed pictures. J. Nerv. & Ment. Dis. 133: 479–488.

Shevrin H, Fisher C (1967) Changes in the effects of a waking subliminal stimulus as a function of dreaming and non-dreaming sleep. J. Abnormal Psychol. 72: 362–368.

Shevrin H, Dickman S (2003) The psychological unconscious: a necessary assumption for all psychological theory? In: Baars BJ (Ed) Essential sources in the scientific study of consciousness. The MIT Press, Cambridge, London.

Smith E, Medin D (1981) Categories and Concepts. Harvard Univ. Press, Cambridge, MA.

Solms M, Saling M (1986) On psychoanalysis and neuroscience: Freud's attitude to the localisationist tradition. Int. J. Psychoanal. 67: 397–416.

Solms M, Turnbull O (2002a) The brain and the inner world. An introduction to the neuroscience of subjective experience. Karnac, London – New York, pp 115–119.

Solms M, Turnbull O (2002b) The brain and the inner world. An introduction to the neuroscience of subjective experience. Karnac, London – New York, pp 101–104.

Spitz RA (1965) Vom Säugling zum Kleinkind (11.Auflage, 1987). Klett-Cotta, Stuttgart.

Sulloway F (1979) Freud: Biologist of the mind: Beyond the psychoanalytic legend. Basic Books, New York, pp 361–415.

Tarpy R (1977) The nervous system and emotion. In: Candland DK, Fell JP, Keen E, Leshner AI, Tarpy R, Plutchik R (Eds) Emotion. Brooks/Cole, Monterey, California.

Tinbergen N (1951) The study of instinct. Oxford University Press, London.

CHAPTER 5

The Microgenesis—A Hierarchical Theory of Mental Function

5.1 The Theory of Microgenesis

Theories about human experience and behaviour are usually based on a few basic functional principles. While behaviourism views mental and behaviour-related phenomena from the perspective of a stimulus-response scheme, cognitive science primarily sees the individual as an information-processing being. In contrast to behaviourism, which only defines the processes that occur between stimulation and response as a process within a "black box", cognitive science also makes inner psychic processes (i.e. thinking, feeling or motivations) its research subject. The cognitive processes suspected in this context are conceived as "information processing". Analogous to the stimulus-response scheme of behaviourism, in the cognitive model an "input" occurs, which leads to an "output" after processing. Neural networks are assumed as the basis for cognitive performance.

The current cognitive neuroscience, which inherently contains cognitivistic concepts, is accused of neglecting the evolutionary view of mental and behaviour-related phenomena. The terms of cognitivism, some of which are borrowed from information technology, imply a brain model that consists of individual modules, similar to the principles of a computer. The function of a computer, however, is independent of the order of assembly of the individual parts, as long as all modules necessary for functionality are present. From a cognitivist perspective, this principle is also

G. Wiest, *Hierarchies in the Brain, Mind and Behaviour*,
https://doi.org/10.1007/978-3-031-78030-1_5

applicable to brain function, i.e. the fact that this is an evolutionarily developed biological system is not given any significance.

The so-called microgenetic theory, in contrast to cognitivism, incorporates evolutionary aspects into its concept of brain function. The term microgenesis originally refers to the Gestalt-psychological term of "actual genesis" (Sander 1928). Actual genesis describes the current emergence of a Gestalt (i.e. a form) or the course of gestalt formation. With a series of experimental approaches, the Leipzig School of Gestalt psychology attempted to investigate the lawful and qualitatively different stages of actual genesis of various mental modalities. According to the laws of actual genesis, a perceived Gestalt does not appear immediately; rather, the perception process first goes through several stages. At the beginning of this process is the perception of a so-called Gestalt framework, followed by a "pre-Gestalt". Only at the end does it come to the actual and constant perception of a "stable" Gestalt (Sander 1930). Microgenesis was introduced by the psychologist Heinz Werner as its own theoretical concept (Werner 1940a). Werner's primary scientific interest was in the experimental exploration of the assumption that development represents a process that starts from undifferentiated structures and leads to the formation of hierarchical structures. This concept is also linked to one of the most important principles of the microgenetic theory, namely the assumption that the unfolding process of every mental event, from perception to thinking, goes through the individual stages of brain evolution and ontogenesis. The microgenetic theory therefore understands every form of human activity, be it perception, thinking, acting, etc., as an "unfolding process". This is based on sequences of developmental steps (Werner 1956). The unfolding process thus equates to an instantaneous evolutionary development that takes place within fractions of a second. The becoming aware of mental events represents the endpoint of several preconscious stages that form the cognition process (Hanlon 1990). Contrary to the concepts of cognitive neuroscience, which view mental processes as the result of the interaction of several cortical modular units, mental states according to the microgenetic theory are the result of a process that follows evolutionarily predetermined hierarchical structures, i.e. is vertically directed (Brown 1998). This implies that the microgenetic process always runs from depth to surface and cannot take place in the opposite direction. Some authors consider so-called event-related potentials (ERP)—i.e. electrical activity that can be derived at the brain surface in the context of cognitive processes—as the neural substrate of microgenesis (Fig. 5.1).

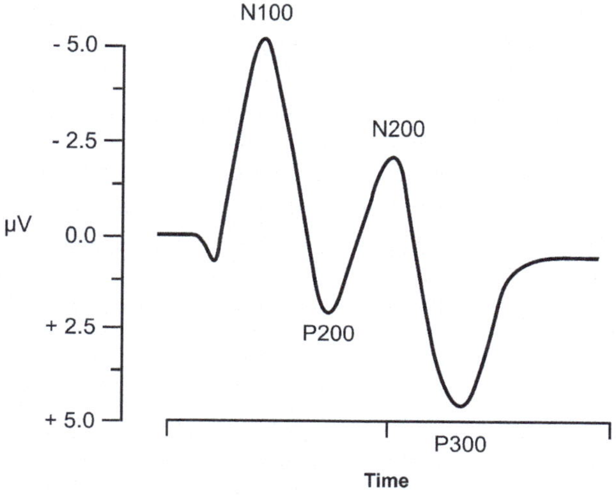

Fig. 5.1 Event-related potential (ERP) with representation of the individual components (N100–P300)

According to the microgenetic theory, a mental event is not to be seen as a result appearing on the surface, but it represents the entire sequence of different unfolding stages. Due to the fact that in the microgenetic process the evolution or phylogeny is recapitulated, it results parallels to psychoanalytic theory, which, however, assumes that the phylogenetic recapitulation occurs in the individual within the framework of his psycho- or ontogenetic development and not during a mental event within milliseconds. Another similarity with psychoanalytic theory is shown in the so-called regression hypothesis of microgenesis. This hypothesis states that any disturbance in the microgenetic process is associated with an exposure of cognitive precursors. Unlike psychoanalysis, which understands regression as a pathological return or fixation at an earlier ontogenetic developmental stage, a disturbance of microgenesis leads to a microstructural regression. This means that symptoms of a microgenetic disorder represent certain levels within the cognitive process and not stages of an individual's cognitive development (Werner 1940b). Otto Pötzl was able, in experimental investigations, long before the introduction of the concepts of actual genesis or microgenesis, to demonstrate a connection between perception processes and unconscious information processing (Pötzl

1917). Using tachistoscopically (i.e. a presentation of images below the perception threshold) generated stimulation images, Pötzl was able to show that image elements that could not be consciously perceived and therefore remained in the Unconscious, subsequently partially returned in modified form in dream images. These results can be seen as experimental confirmation of the importance of "day residues" for dream events postulated by Freud. Similar phenomena could also be induced by Pötzl in patients with perception disorders (agnosias). Such patients described a previously unnoticed image detail in a subsequent image description, i.e. even in this situation, an initially unnoticed (unconscious) object remained in the unconscious and returned in a new context. Pötzl's investigations not only provide an experimental insight into the process of object perception but also confirm the connections between unconscious information processing and early stages of object perception.

In addition to psychoanalysis, the neurological doctrine of Hughlings Jackson also has connections to microgenetic theory. The neurological principle postulated by Jackson is based on a hierarchical organisation of the nervous system, whereby deeper structures are determined by automatic processes, and superficial or more complex structures by voluntary functional principles. As a result, a function is not transferred to a newer structure during the course of evolutionary development but is represented in the new structure in a different way. Jackson referred to this as a "re-representation" (Jackson 1931). The theory also assumes that the higher structures control or inhibit the lower ones. A loss of the higher centres would therefore be associated with a disinhibition and manifestation of the lower systems. From the perspective of microgenesis, early phases of a cognitive process are being incorporated into later phases. This model is thus less characterised by a control of the deeper elements by the surface, but the microgenetic process proceeds from the depth to the surface. Archaic forms of cognition are therefore not seen as disinhibition phenomena but as early manifestations of a developing process. Disturbances of microgenesis, in the form of neurological or psychiatric symptoms, could in this respect be understood as elements of an interrupted process, which have prematurely reached the "surface". The hierarchical theories of Jackson and Freud are—for the reasons mentioned above—consequently designed according to a "top-down" principle, while the process of microgenesis also follows hierarchically organised structures, but does so according to the "bottom-up" principle.

By observing the symptomatology of neurological patients, the neurologist Jason W. Brown made a decisive advancement of the theory of microgenesis founded by Werner (Brown 1988a). A significant element of the microgenetic theory is the conception of deficits and symptoms. While the neurological symptoms of a patient are usually seen as deficits, i.e. as a "lack" of certain abilities, microgenesis focuses on the faulty function. This means that in neurology attention is paid to what a patient can no longer do due to his disorder, while from a microgenetic perspective it is important to see what mistakes he makes due to his disorder. These views have clear reference points to Jackson's theory of neurological symptoms, which in his view always consist of negative and positive elements combined. The error or the positive element of a symptom is thus seen as a snapshot of a process that manifests prematurely and thus in an incomplete form (Brown 1988a). An error is therefore implicitly understood as a manifestation of a normal but interrupted process. A disorder or lesion at a certain point in the brain causes the microgenetic process to proceed normally up to the point of the lesion, but the unfolding process from this point on cannot progress. The appearance of the symptom therefore correlates with the level of the lesion. Analogous to the pathological phenomena of aphasias, "physiological" disturbances of speech—such as slips of the tongue in the sense of a Freudian slip, or speaking in a dream—could be presented as manifestations of a transient "relapse" to deeper functional levels of the speech process (Brown 1988a).

There are several models regarding the pathomechanism of symptom formation. In addition to the already mentioned concept of interruption of the microgenetic process by a lesion—here the symptom would be a premature manifestation of an incomplete process—according to Brown it would also be conceivable to use the principle of heterochrony for the development of symptoms. The term heterochrony refers to different temporal sequences of processes that are subject to the same developmental processes. The mechanism is comparable to a congenital organic developmental defect in childhood, which, despite its morphological deformation, undergoes the same maturation processes as the healthy parts. If one tries to apply this concept to a brain lesion and the microgenetic process, a picture emerges that reminds one of an obstruction in a river. The flow is indeed delayed by the obstacle, but not completely blocked (Brown 1988a).

From the perspective of heterochrony, a symptom points to that brain region where the lesion causes a delay in the "cognitive flow". The image of a "cognitive flow" reminds us of neuroscientific concepts that draw on field theories as the basis of mental phenomena. The ideas of flows or fields contrast with the cognitivist concepts of modules that are connected by pathways. A pathomechanism of symptom formation, based on the principle of "delay" and not "interruption" of the microgenetic process, would also mean that a dysfunction would never be complete.

5.2 Microgenesis and Mental Function

Language

The analysis of language disorders, especially of aphasias, formed the basis for the further development of the microgenetic theory according to Heinz Werner. The current concepts of aphasiology are still influenced by the localisation principles of neurology from the previous century. According to these theories, the ability to speak is ensured by the interaction of individual cortical functional systems (modules). The modules include a "motor" centre (Broca's area), which is located in the anterior frontal lobe, as well as a "sensory" area (Wernicke's area), which is located in the posterior and superior part of the temporal lobe. Language is conceived in these ideas as a process that takes place exclusively within or in interaction between these cortical modules.

Some arguments speak against such a localisation functional principle of language. On the one hand, the modular concept of aphasias with their aetiological classification into motor, sensory and conduction aphasia does not do justice to the diverse clinical appearances of aphasia; on the other hand, temporal and dynamic components of the symptom of aphasia are ignored.

These inadequacies of the current aphasia concepts prompted Brown to examine the symptoms of language disorders from the aspect of a process theory, the microgenesis (Brown 1972). It is interesting to note in this context that Freud also spoke out against the localisation theories and for a dynamic concept of aphasias in his aphasia study (Freud 1891). A dynamic perspective also formed the basis for the development of psychoanalysis (Solms and Saling 1986).

The possibility of understanding language as a development process and aphasia as disruption of this process contradicts common notions that assume predefined modular structures and functions of defined language centres. The observation of the changeability of aphasia symptoms in the course of the developmental process of children provides convincing arguments for an underlying dynamic process (Brown 1978). It has been shown that lesions in the Wernicke language centre at different ages are associated with different forms of aphasia. This phenomenon provides not only insights into phylogenetic and ontogenetic aspects of lateralisation of brain functions but also initial approaches to a microgenetic theory of language and aphasia (Brown and Jaffe 1975).

To understand the microgenetic theory of language and its disorders, it is necessary to consider the evolutionary development of the forebrain. From a microgenetic perspective, the neural substrate of language corresponds to a hierarchical system, the surface of which consists of an anterior and a posterior component. Another factor that is significant in the microgenetic process is the lateralisation of the language function that takes place during ontogenesis. Anterior and posterior components develop in parallel and thus complement each other in the unfolding speech act (Brown 1979a). According to the principle of microgenesis, the language process is to be understood as a process that develops hierarchically across various anatomical strata or levels. Disorders in this sequence provide insight into the microgenetic nature of language. Symptoms of aphasia represent a premature interruption of microgenesis, thus being precursors of the physiological language process. Such an interruption can affect anterior, posterior or both components. The characteristics of the symptoms point to the specific level at which the disorder occurs.

The anterior area of language is due to its specific executive function associated with aspects of action . The microgenesis of the motor language component proceeds according to the microgenetic process from depth to surface across various hierarchical-evolutionary stages of the forebrain. Starting from basal brainstem regions, the microgenetic process continues over the upper brainstem and the basal ganglia. In these regions, so-called pre-programmes for actions are generated, which include aspects of vocalisation, as well as body and limb motor skills. The process continues to mesial parts of the limbic cortex, which is associated with an increasing fragmentation of the aforementioned motor functions. A lesion at this stage leads to disorders of the initiation of vocalisations or extremity movements, to the so-called transcortical motor aphasia or to the "alien hand" syndrome. This is characterised by one hand performing motor actions that are not subject to voluntary control. In addition, such lesions can lead to the occurrence of vocal and motor automatisms (Brown 1978). The further course of the microgenesis of a speech utterance leads to the precentral and premotor cortex (the Broca area) of the dominant hemisphere. In these areas, the structuring of temporal sequences of vocalisation and articulation takes place. Accordingly, disorders at this level lead to phonetic and articulatory deficits. The microgenetic process of language thus passes through the entire evolutionary hierarchy of the cerebral cortex, from the limbic cortex to the focal neocortex. This process applies to both the anterior and the posterior structures of the language function (Brown 1988a).

The posterior component of the language function, analogous to the anterior component, also builds on a hierarchical structure. However, the posterior language area represents not a motor but a perceptual system. The perceptual component of the language function arises, according to the microgenetic theory, in the upper brainstem area; subsequently, it gains additional conceptual and affective content in the limbic cortex. At this stage of the microgenetic process, it is not yet determined whether the perceptive function will be used for language production or language perception. A lesion at this level leads to unusual word confusions—reminiscent of schizophrenic neologisms—and to confabulations (Brown 1979a). As the process continues, there is increasing filtering across several semantic levels. However, a perceived word is not extracted from a memory but rather gains its meaning by considering all components that are part of its microgenesis. The perception of language with all its affective, associative and symbolic components finally takes place in the posterior area of the dominant hemisphere. Lesions at this level lead to symptoms of word confusion and to word-finding disorders (Brown 1988a). The application of the microgenetic theory to language function and its disorders provides explanatory models for aphasia symptoms that are not guaranteed by conventional linguistic and neurological concepts. The ideas about the microgenesis of language are not just theoretical constructs but are based on numerous clinical observations of the manifestations of aphasias (Perecman 1985; Brown 1979b, 1980, 1988a).

Action

An action, according to the microgenetic theory, is also to be imagined as a process of unfolding. Following the hierarchical-evolutionary principle of microgenesis, phylogenetically older structures of motor skills are responsible for early stages of the unfolding process. Accordingly, the microgenetic process of an action develops from regions of the brainstem and the basal ganglia. At this early stage, where connections to perceptual components and the respiratory system already exist, the control and organisation of body posture and movement primarily takes place. In lower animals, such as fish or reptiles, whose motor appearances are limited to trunk control and some elementary movement patterns, this level already represents the end of the microgenetic action process (Brown 1988a). In higher animals, the process continues over mesial paralimbic or frontal areas. At this level, the motor control of proximal limb parts and the organisation of instinctive actions, such as escape reflexes, already take place. The unfolding process experiences further differentiation in the premotor cortex areas, ensuring the control of the distal limb sections, and ends in the precentral region of the neocortex (Brown 1988a).

The unfolding process thus goes through the entire evolutionary development of the motor system, from simple trunk control to limb motor skills to the fine motor skills of the fingers. However, this process is not only inherent in the development from simple to complex, but it also reflects the dynamics of the organisation of motor control. In this respect, bilateral control of actions takes place at the lowest level, followed by a level characterised by the development of dominance and a temporary asymmetric control. Finally, at the highest level, there is a reorganisation with now contralateral action control by the precentral motor cortex. Impressive evidence for the different organisation of motor skills at the various hierarchical levels is provided by motor phenomena in the context of epileptic seizures. While the focal-motor seizures described by Jackson originate from the contralateral motor cortex, the motor automatisms occurring in the context of temporal lobe seizures—such as fidgeting movements—indicate an ipsilateral seizure origin (Chee et al. 1993). This would mean in correlation with the microgenetic theory that motor control at the level of the medial temporal lobe is still bilateral and at the cortical level is already organised only unilaterally and contralaterally.

An interesting phenomenon provides clues for a possible connection between the microgenesis of action and language: A severe damage to the left hemisphere can lead to a complete aphasia and paralysis of the right extremity. Such patients are usually unable to speak or write with their right or left hand. After the creation of a special arm prosthesis for the right arm, which allowed these patients to regain a rough motor ability to write using the preserved proximal motor skills of the shoulder, it was shown that they were able to write words and even complete sentences flawlessly despite complete aphasia (Brown et al. 1983). These results show that only pronounced aphasias and motor deficits allow access to subordinate levels of language organisation. The idea of a sudden "release" of subcortical forms of language organisation is supported by the observation that these patients are not aware of their "regained" abilities. Similar to the sensory phenomenon of blindsight—the ability for implicit visual perception despite damage to the entire visual cortex—these phenomena suggest that actions, like perceptions, do not originate in the primary neocortex but develop in the form of an unfolding process across deeper evolutionary stages.

For Brown, the microgenetic unfolding process also represents another development, namely a transition from an inner to an outer context, i.e. to a goal-directed action, which represents a turn towards the outer world. This means that at the beginning there is a bilateral, "egocentric" control of trunk motor skills, which in the course of development leads to an asymmetric and differentiated control and thus to the "grasping" of the outside world. The microgenesis of action thus goes hand in hand with the microgenesis of object formation or perception (Brown 1988a).

Oscillatory programmes are suspected as the neural substrate of the microgenesis of actions, which initiate and maintain the microgenetic process (Schepelmann 1979). Disturbances in the microgenesis of an action lead to corresponding symptoms, which represent the functioning of the process at the level of the lesion. Consequently, interruptions at the beginning of the process, i.e. at the upper brainstem level, lead to coma or so-called akinetic mutism. Akinetic mutism is a condition characterised by a lack of voluntary actions or reduced action drive, without the presence of a definitive paralysis (Ackermann and Ziegler 1995). According to the microgenetic concept, the origin of the action process is damaged in this syndrome. As a consequence of the disturbance, any initiation of body movements, even at an instinctive level, is impossible. The patient thus not only regresses in terms of motor skills but also in terms of perception to a pre-object stage (Brown 1987). A disturbance of the process further rostrally, in the area of the basal ganglia, leads to specific deficits in trunk and limb motor skills and to clinical manifestations, such as Parkinson's disease. The motor organisation at this level, from the perspective of microgenesis, is not yet oriented towards the capture of an external object (Yakovlev 1948). The dysfunction at this stage of the unfolding process is not only limited to the action level but also includes cognitive and affective components. The connections of the microgenesis of actions and affects are also reflected in lesions in the area of the mesial frontal lobe. Clinically, bilateral disturbances in this area can lead to symptoms that resemble those of akinetic mutism. Before the motor cortex, the microgenetic process reaches the supplementary motor area (SMA). The functional correlate of this region, which is primarily associated with the initiation of movements, is the so-called readiness potential (Kornhuber and Deecke 1965). This electrophysiological phenomenon—a negative potential wave that can be derived from the EEG—can be registered over the SMA just before a motor action. According to the theory of microgenesis, this region is involved in the preparation of actions before they are conscious. A lesion of the SMA is therefore associated with disturbances in the early phase of action initiation. Since at this microgenetic stage of motor skills there is no differentiation of movement patterns yet, a disturbance of the SMA not only includes elements of trunk and limb motor skills but also motor components of speech. Clinically, disturbances of the initiation of both movements and speech can be detected in SMA lesions (Brown 1977).

Bilateral damage to the frontal lobes leads to a complex neuropsychological symptomatology, which in its entirety is referred to as "frontal lobe syndrome". The clinical picture ranges from symptoms of hypo- or hyperactivity, disturbances of drive with apathy or impulsivity, perseverations, attention deficits to paraphasias and apraxias. The heterogeneity of the symptoms associated with the frontal lobe syndrome makes a unified neuropsychological explanation of the clinical picture difficult. Through a microgenetic approach, it is possible to understand the various deficits as a disturbance of the unfolding process of actions. The different qualities of the symptoms can ultimately be understood by the interruption of the action process at different evolutionary levels. An interruption at the lowest level in the area of the frontal lobes manifests itself in the inability to start an action (apathy, mutism) or to switch between two action options (perseverations). Disturbances at a higher level of the process lead to impairments of an action after its initiation in the form that there are dysfunctions in the sequence of actions. This symptomatology can range from general action-related phenomena, such as increased impulsivity or distractibility, to specific disturbances of the action sequence (apraxia). Damage at the end of the microgenetic process finally leads to deficits at the last stage of the execution of actions. Accordingly, there are disturbances of motor execution in the form of paresis in the area of the limbs, articulation disorders, as well as restrictions of motor aspects of speech (Brown 1987).

Perception and Memory

Current neuroscientific concepts assume that perception represents a process that begins with a specific stimulation at the sensory organ. The information is subsequently transmitted via specific pathways to ultimately lead to the perception of the object at the corresponding sensory cortex. A comparison of the current perception with previous experiences is made via connections to other structures and cortical areas. In this way, representations of certain objects are formed.

In contrast, the microgenetic theory does not understand the formation of an object representation as a process characterised by the assembly of sensory information. According to these views, perception is also not an event that primarily takes place at the cortical level. Rather, perception from the perspective of microgenesis is understood as an unfolding process in which representations form along evolutionary stages of brain development. The process is thus conceived as a process that runs parallel through sensory and mental levels, corresponding to the hierarchical structure of the forebrain (Brown 1988a). The individual levels of perception implicitly correlate with the levels of object representation. The final stage of the perception process consists of the sum of all previous representation levels. Correlating with the basic principles of microgenesis, an interruption of the unfolding process at certain levels leads to specific pathological appearances. The resulting symptom of a lesion corresponds—analogous to the disorders of language or action—to a prematurely terminated unfolding process. Each damage leads to symptoms that either go hand in hand with a functional deficit (such as a loss of vision) or give rise to abnormal sensory appearances (e.g. hallucinations). The described symptomatology can therefore manifest itself in the form of negative (deficits) or positive (abnormal appearances) phenomena (Brown 1983a).

The microgenesis of the perception process was primarily designed for the visual system; however, the model can also be transferred to other sensory systems (Brown 1988a). According to the microgenetic theory, object perception begins in the area of the upper brainstem and the tectum as a two-dimensional representation of the body surface. This ponto-mesencephalic level of sensory processing forms the basis for all perceptual qualities and is also considered by some authors as the neural substrate of dreamless sleep. The mental function of dreaming is only ensured by limbic structures according to these ideas (Hernandez-Peon 1966).

At the ponto-mesencephalic level of the perception process, the predominantly body-related sensory processing prevents the capture of external objects, so that the representations of body and external object are not yet differentiated at this stage, i.e. the spatial concept of body and external world still coincide at this level. The impossibility of capturing an external object independent of the body means that objects can only be perceived as targets. A pronounced damage to the ponto-mesencephalic region can lead to coma or akinetic mutism, in addition, disturbances of the sleep-wake cycle and hallucinations can be observed (Brown 1988b). The hallucinations associated with upper brainstem lesions, also known as "peduncular hallucinations", are among the so-called hypnagogic phenomena. The appearances are predominantly of a visual nature, but auditory or haptic sensations are also possible. The term hypnagogic refers to the characteristic that the hallucinations mostly occur in the transition phase to sleep. In contrast to other forms of hallucinations, hypnagogic phenomena are usually simply structured and not associated with any affect changes (Brown 1985).

A further transformation of the microgenetic perception process takes place in the limbic system or in the temporal lobe. At this level of perception, an association of sensory information with instinctive, affective and mnestic components takes place, which gives the object representation an egocentric dimension. Pictorial sensory information can also have hallucinatory properties at this stage. Accordingly, this microgenetic level of the perception process is also represented in normal dream events. In this context, it would be conceivable that the sequence of dream and dreamless sleep is reflected in the interplay of brainstem and temporal lobe mechanisms (Brown 1988a).

Lesions in the temporal lobe can be associated with perceptual deficit symptoms, in the form of agnosias (disorders of "recognition"). The agnosias, which are associated with mesial temporo-occipital damage, show as a main feature mnestic deficits and the associated problems in capturing the meaning of objects. The latter is associated with an inability to establish symbolic and categorical connections to objects. The symptoms of agnosias, which occur in parietal or parieto-occipital lesions, on the other hand, are understood as a disorder of object identification or discrimination, and represent a spatial perception deficit (Brown 1983a).

In contrast to the relatively simple hallucinations in lesions in the upper brainstem, the complex hallucinations of the temporal lobe are characterised by condensation and symbolic transformation; in addition, they often have a pronounced affective tone. Further characteristics of temporal hallucinations are their reference to the personal experiences of the perceiver, as well as their scenic properties, i.e. they are not rigid images or appearances but complex temporal sequences of experience content (Brown 1985).

The perception of an object as an independent, external entity in a three-dimensional space is only achieved at the cortical level in the area of the parietal lobe. At this stage of the microgenetic process, the transformation from an egocentric space concept to a more object-related context also takes place. Lesions in the parietal lobe consequently lead to symptoms that are associated with a disturbance of the spatial perception of objects. The latter means that objects are perceived as changed in terms of their size, orientation or position. In perceptual terms, so-called illusory misidentifications are also possible at this cortical level, which manifest as changes or distortions of the perceived objects. A reduced sensory input at the cortex is associated with an autonomous functioning of the structure, which—in contrast to the temporal lobe—is not associated with hallucinations but with image ideas. Changes that on temporal lobe level lead to hallucinations and dream-like images, manifest themselves on the parietal cortical level as altered memory images (metamorphopsias). The disturbance at the cortical level is thus more "object-oriented" than the dysfunction at the evolutionarily deeper limbic level.

The microgenesis of perception ends in the sensory cortex (for vision, for example, in the striate cortex), where the actual recognition process of an external object in its entirety takes place. At the same time, representations of objects, which are perceived as independent entities of the outside world, arise here at the mental level. Physiologically, various phenomena can be assigned to this level of the perception process. These include, in the visual system, phenomena such as afterimages or phosphenes (visual stimulus symptoms after cortical stimulation). Eidetic images are also attributable to cortical mechanisms. An eidetic image is to be understood as a memory trace of a perception just experienced. Disorders at this highest level of the perception process manifest themselves, for example in the case of an interruption of visual information, in the form of elementary hallucinations. Lesions in the striate cortex, on the other hand, lead to vision losses (visual field defects) or to specific visual failures, in the form of colour or shape perception disorders (Brown 1988b). According to the microgenetic theory, perception represents an unfolding process of sensory information that runs through various hierarchical stages of brain evolution. Parallel to this process, mental representations of objects develop at different phylogenetic levels of the brain, leading to a differentiated perception of the outside world at the cortical level (Brown 1988a). Memory can be seen as a mental performance based on previous perceptions and experiences. Taking into account the broad spectrum of mnemonic functions, memory performances in the models of cognitive science are divided according to various criteria. Among other things, a distinction is made between long-term and short-term memory, or between explicit and implicit memory. Current neuroscientific concepts also assume that specific memory functions can be assigned to certain neural structures. These ideas correlate with the image of a modular working method of the individual memory systems, whereby certain information can be stored and retrieved in the corresponding brain areas.

Starting from the question of what distinguishes the memory of an object from the recognition and re-recognition of an object, the microgenetic theory seeks a different explanatory model for memory. In this respect, perceptual phenomena such as visual afterimages or déjà-vu experiences could be understood as intermediate stages of a mnemonic or perceptive continuum. The microgenesis of memories would not have to be seen as a separate unfolding process but could be linked to the microgenetic concept of perception and object formation. According to Brown's theory, an idea (or a memory) represents an object that has undergone the microgenetic unfolding process only incompletely (Brown 1988a). Accordingly, a memory would be a stage in the microgenesis of perceptions or in the formation of object representations. Depending on the extent to which the microgenetic unfolding process of object formation has already progressed, an event is experienced as either a mnemonic or perceptive phenomenon according to this theory. If the unfolding process of a representation reaches the stage where the perception of an object in external space is possible, the object is perceived. A limitation of the process to the aforementioned stage, on the other hand, does not lead to real object perception but only to an idea of the object.

With the model of memory function as a process that occurs along evolutionary levels of brain development, new aspects arise in the understanding of specific mnemonic functions. In principle, from the perspective of microgenesis, visual afterimages, which appear immediately after viewing an object after the eyelid closes, are to be interpreted as mnemonic phenomena. The vivid experience of the afterimage of a just perceived object suggests that here again almost the entire unfolding process of the respective object representation was repeated, i.e. that the memory image comes very close to the actual perception image (Brown 1988a).

The so-called short-term memory, i.e. the ability to recapitulate contents a few seconds to minutes after perception, can also be understood as a manifestation of a microgenetic process. Compared to a visual afterimage, performances of short-term memory involve a genuine mental reconstruction of a previously made perception or experience, so it is more of a cognitive than a perceptive process. Memories from short-term memory represent, in contrast to afterimages, earlier stages of object formation, i.e. levels in which the ability to perceive an external object has not yet developed. As a consequence of this fact, the object reconstructed by the short-term memory already lacks numerous characterising elements (Brown 1988a).

Analogous to the theory of short-term memory, the microgenetic concept of long-term memory is based on similar principles. The reconstructions from long-term memory are characterised by their rudimentary nature. The memories are usually retrievable in the context of other events, with temporal sequences hardly taken into account anymore. The contents that are recapitulated from long-term memory thus come from the earliest stages of the perception and object formation process.

The disorders of memory referred to as amnesias offer new possibilities for interpreting the mnemonic function from the perspective of microgenetic theory. According to Brown, the symptoms of amnesia are not due to a deficit in retrieving content but are an expression of an incomplete unfolding process in object formation (Brown 1988a). From a microgenetic perspective, the symptom reflects the normal act of remembering up to the level of interruption (or delay) by a lesion, with certain associated elements of memory possibly remaining depending on the level of the disorder. In such a case, it is not possible to remember a specific content, but associated thoughts or affects come to consciousness. A well-known observation by Eduard Claparède provides an example of this: He greeted amnesic patients, who did not remember previous encounters with him during the visit, one day with a needle in his hand. The patients did not recognise the doctor the following day either but refused to shake his hand in greeting (Claparède 1911). This observation illustrates that in such cases an explicit memory is not possible, but an implicit memory can indeed influence behaviour. The phenomenon can be seen as a manifestation of an incomplete microgenetic unfolding process, in which only associated components of the actual microgenesis persist. A similar interpretation would be possible for the symptom of "confabulation" in patients with Korsakoff's syndrome. In this case, the amnesia for certain contents is compensated by confabulations, i.e. imaginative narratives with associative character are "filled in". Thus, confabulation becomes the manifest (residual) symptom of an impaired unfolding process (Brown 1988a).

Consciousness

The principle of microgenesis is not only applicable to various forms of mental functions but also forms the basis for a theory of consciousness (Brown 1983b). The microgenetic unfolding process is conceived as a process that develops along evolutionarily developed brain structures. The direction of the process is strictly unidirectional, along a vertical axis (caudal-rostral), and thus not reversible. The microgenetic theory of consciousness assumes that an unfolding process that spreads across evolutionary levels leads to the development or emergence of new properties. This phenomenon is referred to in biophilosophy as "emergence" (appearance). The concept of emergence goes back to the theories of the philosopher G. H. Lewes, who differentiated between resultant and emergent properties (Lewes 1879). In this respect, a property of a whole that already possesses one of its components is referred to as "resultant". However, if a whole has a property that none of its parts possess, it is called "emergent" (Mahner and Bunge 2000). The property of being alive can be seen in this context as an emergent property of a cell. Similarly, the ability to be conscious would be seen as an emergent property of the brain.

In relation to a microgenetic theory of consciousness, the question arises as to how the emergent properties in the sequence of an unfolding process are to be imagined. In the case of a segmental structure of the microgenetic process—in which each segment could be assigned to a corresponding evolutionary level of the brain—the transition from one level to the next would have an emergent character. In another model, the unfolding process would not be divisible into individual few segments but due to the smallest segmentations would be seen as a continuum. The idea of a continuum would mean that states of consciousness would be the result of a succession of constantly changing segments (Brown 1983b).

By linking the microgenetic theory with data from psychophysiology, it becomes possible to create a concept of the immediate and conscious experience of the "now". Psychophysiological models assume a "moment of perception" that forms the basis for the continuous experience of a "present". This results from the sequence of shortest perception spans, each lasting approximately 0.1 seconds (Stroud 1967). From a microgenetic perspective, every unfolding process that extends from deep brain regions to the cortical endpoint can be assigned to such a perception moment and thus forms the substrate for the experience of the "now" (Brown 1983b). Only through the combination of sequence and superimposition of microgenetic processes over time does one experience state replace the next, i.e. every conscious experience develops overlapping with the preceding one. From the subjective experience state of the "now", which lasts only fractions of a second, there is a flowing transition to the contents of short-term memory, which correlate with the slow decline of the superimposed microgenetic processes. With increasing temporal distance and corresponding reduction of processes, the transition from contents to long-term memory follows (Brown 1988a).

For the occurrence of an experience of the "now", the unfolding of a microgenetic process is required, which develops from archaic limbic structures towards the various evolutionary stages of the neocortex. Thus, earlier levels of the perception and object formation process can be assigned to evolutionarily older neural structures. At the same time, the various neural and microgenetic levels correlate with certain memory functions, i.e. short-term and long-term memory.

The idea, based on psycho-physical data, that consciousness could arise through repetition of shortest perception moments or spans at a frequency of 10/sec, allows for the assumption of a "biological generator" as a neural substrate (Brown 1983b). This would be obvious , as such oscillators have already been detected in several brain areas. To mention in this context are oscillators in the upper brain stem, which are associated with aspects of the sleep-wake cycle, or those that are connected with the motor or respiratory system. According to Brown, it would be possible for rhythms at different levels of the microgenetic process to differentiate from a basic rhythm and thus form the basis for conscious experience (Brown 1982).

The microgenetic approach opens up new aspects in the understanding of altered states of consciousness. The concept of attention is understood from the perspective of microgenesis as a behaviour-related manifestation of perception. Attention appears implicitly to be associated with a focal orientation towards an object. It thus corresponds to a specific cortical function and represents the endpoint of a microgenetic process. The paradoxical phenomenon of a "diffuse" attention would therefore be compatible with a perception process at a lower neural developmental stage, i.e. with an incomplete microgenetic process. Indeed, the symptom of confusion can be seen as a manifestation of an attention disorder in the sense of a return to a diffuse attention (Brown 1983b), i.e. a state of confusion also corresponds to an early form of perception and cognition. With the emergence of preliminary components of the microgenetic attention process in the form of confusion states, there are often other symptoms associated such as hallucinations or affective disorders. Confusion also seems to be associated with a profound disturbance of the sense of time, which ultimately also impairs the ability to perceive the "now" (Brown 1983b).

5.3 Microgenesis and Psychopathology

The unfolding processes of cognition and perception form the basis for a microgenetic theory of psychopathological disorders. The concept of microgenesis understands every form of development as a process that leads from a general and undifferentiated state to a differentiated and hierarchically integrated state. In terms of thinking, this means that the basic and undifferentiated stage of thought organisation can be presented as an accumulation of images or simple concepts. These early forms of thinking are also referred to as "primary aggregations" of related ideas in microgenetic theory (Arieti 1962). The "primary aggregations" are organised in such a way that at this level, rather than specific contents, semantic connections are captured. Disorders of thinking can be observed in patients with aphasia or schizophrenia. From a microgenetic perspective, the act of thinking in these patients appears to be fixed at an early stage. The categorical confusions of concepts often described in schizophrenic patients, as well as the so-called semantic paraphasias of aphasia patients, can be seen as manifestations of an early organisational stage of thinking. The question of the name of the Pope can be answered by a schizophrenic patient with the word "Vatican". The word found does not fall into the category of papal names but is closely related in meaning to the term "Pope". This form of thinking would therefore be attributable to an early stage of cognition, at which no differentiation of specific contents is possible and a loose connection between the individual elements exists.

This mode of operation is reminiscent of the principle of primary process described by Freud. Here too, thought contents are loosely and associatively connected and mechanisms such as condensation and displacement predominate, so that similar elements can easily be exchanged.

The response behaviour of the previously mentioned schizophrenic patient even occurs at the undifferentiated stage of thinking, not randomly, but suggests that even here simple forms of groupings exist. Within these so-called primary groups, the contents are equivalent, as they possess a common characteristic. The term "God" can, for example, be used synonymously for the word "King", as both contents represent special instances. The special property that connects both words suggests that the words originate from a common group. Observations of patients with schizophrenia suggest that their thinking is not fixed at an earlier stage in every respect. Arieti pointed out that this is only the case when thought contents are connected with certain personal aspects of the patient (Arieti 1948). A schizophrenic patient, for example, thinks that she is the Virgin Mary because she herself is virginal like Mary, i.e. here Mary and the patient share a common characteristic. On the other hand, an early form of thinking is only manifested here because virginity has a certain personal meaning or represents a certain psychodynamic problem for the patient (Arieti 1950).

In contrast to the more severe disorder of schizophrenia, which from a microgenetic perspective corresponds to an early stage of thinking, psychoneuroses represent more mature psychopathological phenomena. In neurotic disorders, according to Arieti, there is no identification with objects or contents due to similarities, as in schizophrenia.

The neurotic behaves towards other objects as if he/she belonged to a primary aggregation, i.e. objects and contents that can be attributed to an early stage of thought organisation (Arieti 1962). Arieti gives the example of a neurotic patient who exhibits noticeably aggressive behaviour towards older red-haired women. The causes were not conscious to the patient until he remembered during therapy that his mother often used to take him to his disliked aunt for care, who had a similar appearance. This aunt behaved very dismissively and uninterested towards him. In this case, his original behaviour towards the aunt was transferred to the entire category of older red-haired women. The connection between his current behaviour and the origin of the behaviour remained unconscious to the patient for a long time. Here, a discrepancy is shown between the current cognitive perception of a woman with the described physical characteristics and the behaviour towards her. The patient behaves towards the women as if they were a primary collection of similar objects. This behaviour ultimately also originates from a primitive level of thought organisation. However, this manifests itself in neurosis only in actions that are perceived by the person themselves as alien and inexplicable. In the case of schizophrenic disorders, on the other hand, there is a deeper impairment of cognition or logical thinking.

The question of the reasons for a certain response behaviour of patients with disorders of the microgenetic cognition process can be answered in that personal experiences also play a role in the choice of certain answers.

It is understandable that a schizophrenic patient, when asked about a certain capital, chooses another city, as this also belongs to a primary group. Why his choice falls on a very specific city, can be due to personal experience. Further dynamic explanations for the preference of certain cognitive elements could also be found in unconscious or primitive desires. A patient may, as an expression of denial of his illness, therefore believe he is in a hotel and not in a hospital (Weinstein and Kahn 1955).

Using the example of the neurotic patient, whose original emotional conflict with his aunt remained unconscious and was subsequently repeated at every encounter with similarly looking women, the question arises as to the benefit of a fixation of behaviour at this stage. It can be argued that the repetition of the early forms of behaviour is currently experienced as alien by the patient, but at the same time it serves to avoid displeasure, which would be associated with the painful conscious memory of the aunt.

The previous explanations about the microgenetic unfolding process of cognition reveal some basic principles: Depending on the presence or extent of a disorder, the cognition process can be interrupted or fixed at various levels. A fixation at a certain stage can not only occur through organic damage but can also be caused by motivation or unconscious desires. It can therefore be concluded that the cognition process at each level of microgenesis is accompanied by a corresponding affective change. In qualitative terms, it can be assumed that every thought process before becoming conscious is initially unconscious.

According to Arieti's microgenetic theory, it is primarily the unconscious affect that determines whether a particular stage of the cognition process becomes conscious or not (Arieti 1962).

For the microgenesis of perception, similar principles are assumed as in the microgenetic cognition process. Certain psychopathological phenomena of schizophrenic patients seem to provide insight into early forms of perception. A particular perceptive deficit of these patients is the inability to perceive objects in their entirety or wholeness (Arieti 1961). People are not recognised by the patients as individuals, but only parts of the body are perceived individually, so to speak as a form of perceptive fragmentation. Such phenomena are also described in states of confusion of other aetiology. From a microgenetic point of view, these fragmentations of perceptions would be understood as a return to more primitive stages of the perception process. The individual parts of an object would thus be seen again as "primary aggregations" of perceptual elements. As indirect confirmation of the existence of early forms of perception, the observation of patients with congenital blindness can be used: These patients report after a cataract operation, which for the first time enables them to see that they have only perceived fragments of objects in their first visual impressions. Further evidence for the existence of early stages of perception comes from subliminal research. Pötzl was already able to demonstrate in the visually induced dream images of normal people that subliminally, i.e. below the perception threshold perceived image, in the subsequent dream only appears as a partial object or fragment (Pötzl 1917). These experimental results would be well compatible with the idea that unconscious perceptions correlate with the precursor of microgenetic perception processes.

In summary, it can be said about the microgenesis of cognition and perception that the unfolding process of both mental performances initially goes through unconscious phases before it reaches consciousness. Neuropsychological and psychopathological disorders allow conclusions to be drawn about the existence of microgenetic processes. The microgenetic theory understands psychopathological symptoms as unfolding processes that come to the "surface" too early and therefore incompletely. The incomplete thought processes that come to the surface can now be in line with the behaviour they provoke or not. In the case of congruence between early thought processes and behaviour, a psychosis results, in the case of incongruity according to microgenetic theory a psychoneurosis.

Bibliography

Ackermann H, Ziegler W (1995) Akinetic mutism – a review of the literature. Fortschr. Neurol. Psychiatr. 63: 59–67.

Arieti S (1948) Special logic of schizophrenic and other types of autistic thought. Psychiatry 11: 325–338.

Arieti S (1950) Primitive intellectual mechanisms in psychopathological conditions: study of the archaic Ego. Amer. J. Psychother. 4: 4–15.

Arieti S (1961) The loss of reality. Psychoanalysis 48: 3–24.
Arieti S (1962) The microgeny of thought and perception. Arch. Gen. Psychiatry 6: 76–90.
Brown JW (1972) Aphasia, Apraxia, and Agnosia. Thomas, Springfield, Illinois.
Brown JW (1977) Mind, brain and consciousness: The neuropsychology of cognition. Academic Press, New York.
Brown JW (1978) Lateralisation: A brain model. Brain & Lang., 5: 258–261.
Brown JW (1979a) Language representation in the brain. In: Steklis H, Raleigh M (Eds) Neurobiology of social communication in primates. Academic Press, New York.
Brown JW (1979b) Thalamic mechanisms in language. In: Gazzaniga M (Ed.) Handbook of Neuropsychology, Plenum Press, New York pp 215–236.
Brown JW (1980) Brain structure and language production: a dynamic view. In: Caplan D (Ed.) Biological studies of mental processes. MIT Press, Cambridge, MA.
Brown JW (1982) Hierarchy and evolution in neurolinguistics. In: Arbib M, Caplan D, Marshall J (Eds) Neural models of language processes. Academic Press, New York.
Brown JW (1983a) Microstructure of perception: physiology and patterns of breakdown. Cognition and Brain Theory 6: 145–184.
Brown JW (1983b) Emergence and time in microgenetic theory. J. American Acad. Psychoanal. 11: 35–54.
Brown JW (1985) Imagery and the microstructure of perception. J. Neurolinguistics 1: 89–128.
Brown JW (1987) The microstructure of action. In: Perecman E. (Ed.) The frontal lobes re-visited. IRBN Press, New York.
Brown JW (1988a) The life of the mind: Selected papers. Lawrence Erlbaum Associates, Hillsdale, New Jersey.
Brown JW (1988b) Neuropsychology of visual perception. Lawrence Erlbaum Associates, Hillsdale, New Jersey.
Brown JW (1998) Psychoanalysis and process theory. Ann. N.Y. Acad. Sci. 843: 91–106.
Brown JW, Jaffe J (1975) Hypothesis on cerebral dominance. Neuropsychologia 13: 107–110.
Brown JW, Leader B, Blum C (1983) Hemiplegic writing in severe aphasia. Brain and Language 19: 204–215.
Chee MWL, Kotagal P, Van Ness PC, Gragg L, Murphy D, Lüders HO (1993) Lateralising signs in intractable partial epilepsy: blinded multiple-observer analysis. Neurology 43: 2519– 2525.
Claparéde E (1911) Recognition et moiieté. Archives de Psychologie 11: 79–90.
Freud S (1891) Zur Auffassung der Aphasien. Eine kritische Studie. Fischer Verlag, Frankfurt am Main 2001.

Hanlon R (1990) Cognitive microgenesis: A neuropsychological perspective. Springer-Verlag, New York.
Hernandez-Peon R (1966) A neurophysiological model of dreams and hallucinations. J. Nerv. Ment. Dis. 141: 623–650.
Jackson JH (1931) Taylor J. (Ed) Selected writings of John Hughlings Jackson. Staples Press, London.
Kornhuber HH, Deecke L (1965) Hirnpotentialänderungen bei Willkürbewegungen und passiven Bewegungen des Menschen: Bereitschaftspotential und reafferente Potentiale. Pflügers Arch. 281: 1–17.
Lewes GH (1879) The study of psychology. Houghton, Boston.
Mahner M, Bunge M (2000) Philosophische Grundlagen der Biologie. Springer Verlag, Berlin.
Perecman E (1985) Ukeleles, Condessors, and Fosetch. Language Sciences 7: 177–214.
Pötzl O (1917) Experimentell erregte Traumbilder und ihre Beziehungen zum indirekten Sehen. Zeitschr. Ges. Neurol. Psychiatr. 37: 278–349.
Sander F (1928) Experimentelle Ergebnisse der Gestaltpsychologie. In: Becher E (Ed) 10. Kongressbericht experimentelle Psychologie. Fischer, Jena.
Sander F (1930) Structures, totality of experience, and Gestalt. In: Murchinson C (Ed) Psychologies of 1930. Clark University Press, Worcester, MA, pp 188–204.
Schepelmann F (1979) Rhythmic patterns of motor activity after lesions of the central nervous system in man. Acta Neurochirurgica 49: 153–189.
Solms M, Saling M (1986) On psychoanalysis and neuroscience: Freud's attitude to the localisationist tradition. Int. J. Psychoanal. 67: 397–416.
Stroud J (1967) The fine structure of psychological time. Annals N. Y. Acad. Sci. 138: 623–631.
Weinstein EA, Kahn RL (1955) Denial of illness: symbolic and physiological aspects. Charles C. Thomas Publisher, Springfield, Illinois.
Werner H (1940a) Musical „micro-scales“ and „micromelodies.“ Journal of Psychology 10: 149–156.
Werner H (1940b) Comparative psychology of mental development. Harper, New York.
Werner H (1956) Microgenesis and aphasia. Journal of Abnormal & Social Psychology 52: 347–353.
Yakovlev PI (1948) Motility, behaviour, and the brain. Stereodynamic organisation and neural coordinates of behaviour. J. Nerv. Ment. Dis. 107: 313–335.

CHAPTER 6

Concluding Remarks

The current neuroscientific ideas about neural and mental functions are primarily oriented towards the modular concepts of information technology. One aspect that is largely pushed into the background is the assessment of brain function from an evolutionary perspective. The theory concepts of various scientific disciplines summarised in this book are largely based on evolutionary or hierarchical principles that are applicable to both neural and mental functions.

The beginnings of this perspective go back to the philosopher Herbert Spencer, who introduced an explanatory model for the phylogenesis and the pathologies of the brain with the concepts of evolution and dissolution. Hughlings Jackson found the Spencerian principle of dissolution of the nervous system confirmed in the neurological symptoms of his patients. From this perspective, the possible cerebral failures (dissolution) provide an insight into the phylogenetic development (evolution) of the brain. For Jackson, evolution represents a process from a state of highest to lowest organisation. The validity of the hierarchical organisation of the brain postulated by Jackson is largely confirmed by modern neurological research. Hierarchical organisational principles can be found in the motor and somatosensory system, but also in the visual system.

The observation of homologous behaviour of related species, such as between humans and primates, provides insights into the evolutionary-hierarchical principles of brain development. Innate behaviours in the

G. Wiest, *Hierarchies in the Brain, Mind and Behaviour*,
https://doi.org/10.1007/978-3-031-78030-1_6

form of instinctive movements are found in humans in pure form almost exclusively in newborns or infants, as early ontogeny represents the phylogenetic history at least partially. In adult humans, residues of instinctive actions can still be found in rudimentary form, such as in greeting gestures, which can be attributed to the instinctive submissive behaviour of many mammals. Another possibility to observe instinctive movements or actions in humans is offered—according to the Jacksonian principle of dissolution—by disease processes that are associated with focal cerebral lesions or lead to a diffuse degradation of the brain. Depending on the degree of dysfunction of the brain, these instinctive movements occur reflexively or automatically. In the dissolution of brain function, phylogenetic and ontogenetic stages of development are thus revealed or repeated.

A hierarchical organisation of instinctive actions has already been postulated by comparative behavioural research. However, these views require a corresponding hierarchical structure of functional instances in the brain. The field of research of evolutionary neuroethology attempts to explore the relationships between specific behaviours and brain structure. According to Paul MacLean, the human brain, from an evolutionary perspective, is a "Triune Brain", i.e. an evolutionarily developed three-part neural system, which represents the phylogenetic heritage of reptiles, as well as lower and higher developed mammals. These associations form a unit, corresponding to a conglomerate of three brains—although they differ significantly in their structure and neurochemistry, and are aeons apart in evolutionary terms. Although the associations are highly interconnected, there are also indications for independent functioning. The findings of evolutionary neuroethology also find their way into the ideas of evolutionary psychiatry, especially with regard to the genesis of psychopathological symptoms.

Hierarchical aspects also include the metapsychological theories of psychoanalysis. In addition to the models of the psychic apparatus, the concepts of psychic functioning—the so-called primary and secondary processes—or the phenomenon of regression are based on hierarchical principles. The ontogenetic and phylogenetic references in psychoanalytic theory also involve hierarchical ideas. The field of research of neuropsychoanalysis attempts to investigate or test metapsychological theories using neuroscientific methods. The data from evolutionary neuroethology provide numerous concepts that could be linked with psychoanalytic theory. MacLean's concept of emotomentation and ratiomentation shows here some parallels with the principles of the primary and secondary processes discovered by Freud. Similarly, the pictorial and drive-determined contents of the dream events can be linked to the activity of the limbic system, which is disconnected from the cortex.

The theory of microgenesis assumes that every mental event is the result of an unfolding process that goes through the individual stages of evolution and ontogenesis. This process, which follows evolutionarily predetermined neural hierarchical structures, is applicable to aspects of perception, action and consciousness. A disturbance of the microgenetic unfolding process, i.e. a neurological symptom caused by a brain lesion, is seen as a snapshot of a process that manifests prematurely and thus in an incomplete form. The appearance of the symptom correlates with the level of the lesion.

All the hierarchical concepts presented here, which find their expression in theories of brain function, behaviour and mental function, are based on a top-down approach. This means that pathological phenomena form the starting point for the conception of the hierarchical principle. The arc spans from neurological disinhibition phenomena to neurologically induced atavistic behaviours in humans—and their homologies to animal behaviour—to the psychoanalytic concepts of psychopathological symptoms. Finally, the theoretical ideas of microgenesis are also based on observations of neuropsychological phenomena. The bottom-up approach of neuroscience often provides no conclusive or only insufficient answers to physiological and pathological phenomena of human behaviour. Thus, the aspect of subjective experience is almost completely ignored in current neuroscience, so that there are no coherent concepts for a number of specifically human experiences, such as laughter, crying, behavioural habits, dreams, humour or psychopathological phenomena. In contrast, a hierarchical-evolutionary approach offers not only comprehensive explanatory models for function but also for the disorders of the brain and behaviour.

Bibliography

Ackermann H, Ziegler W (1995) Akinetic mutism – a review of the literature. Fortschr. Neurol. Psychiatr. 63: 59–67.

Adolphs R, Tranel D, Damasio H, Damasio A (1994) Impaired recognition of emotion in facial expressions following bilateral damage to the human amygdala. Nature 372: 669–672.

Adolphs R, Tranel D, Hamann S et al. (1999) Recognition of facial emotion in nine subjects with bilateral amygdala damage. Neuropsychologia 37: 1111–1117.

Adolphs R, Damasio H, Tranel D et al. (2000) A role for somatosensory cortices in the visual recognition of emotion as revealed by 3-D lesion mapping. J. Neurosci. 20: 2683–2690.

Akert K, Gruesen RA, Woolsey CN, Meyer DR (1961) Klüver-Bucy syndrome in monkeys with neocortical ablations of temporal lobe. Brain 84: 480–498.

Albe-Fessard D, Buser P (1979) Is a theory on neostriatal functions now possible?. In: Divac I and Oberg RGE (Eds.) The Neostriatum, Pergamon Press, Elmsford, N.Y., 315–319.

Anschel S (1977) Functional specificity of vocalizations elicited by electrical brain stimulation in the turkey. Brain Behav. Evol. 14: 399–417.

Arieti S (1948) Special logic of schizophrenic and other types of autistic thought. Psychiatry 11: 325–338.

Arieti S (1950) Primitive intellectual mechanisms in psychopathological conditions: study of the archaic Ego. Amer. J. Psychother. 4: 4–15.

Arieti S (1961) The loss of reality. Psychoanalysis 48: 3–24.

G. Wiest, *Hierarchies in the Brain, Mind and Behaviour*, https://doi.org/10.1007/978-3-031-78030-1

Arieti S (1962) The microgeny of thought and perception. Arch. Gen. Psychiatry 6: 76–90.

Arroyo S, Lesser RP, Gordon B, Uematsu S, Hart J, Schwerdt P et al. (1993) Mirth, laughter and gelastic seizures. Brain 116: 757–780.

Auffenberg W (1978) Social and feeding behaviour in Varanus komodoensis. In: Greenberg N, MacLean PD, (Eds.) The behaviour and neurology of lizards. U.S. Government Printing Office, Washington, D.C., pp 301–331.

Auffenberg W (1981) The behavioural ecology of the Komodo Monitor. University Presses of Florida, Gainesville.

Aull-Watschinger S, Pataraia E, Baumgartner C (2008) Sexual auras: predominance of epileptic activity within the mesial temporal lobe. Epilepsy Behav. 12: 124–127.

Baddeley A (1989) Hierarchies and human memory. In: Kennard C, Swash M (Eds) Hierarchies in neurology. A reappraisal of a Jacksonian concept. Springer-Verlag, London, New York, pp 49–54.

Baerends GP (1956) Aufbau des tierischen Verhaltens. In: Helmcke JG (Ed) Handbuch der Zoologie, Band VIII, de Gruyter, Berlin.

Bailey KG (1978) The concept of phylogenetic regression. J. Amer. Acad. Psychoanal. 6: 5–35.

Baumgartner C, Gröppel G, Leutmezer F, Aull-Watschinger S, Pataraia E, Feucht M, Trinka E, Unterberger I, Bauer G (2000) Ictal urinary urge indicates seizure onset in the non- dominant temporal lobe. Neurology 55: 432–434.

Beach FA (1951) Effects of forebrain injury upon mating behaviour in male pigeons. Behaviour 4: 36–59.

Bellairs A (1970) The life of reptiles. Universe Books, New York.

Bingley T (1958) Mental symptoms in temporal lobe epilepsy and temporal lobe gliomas. Acta Psychiat. Neurol. Scand. 33, Suppl. 120.

Bird A (2007) Perceptions of epigenetics. Nature 447: 396–398.

Brakel LAW, Kleinsorge S, Snodgrass M, Shevrin H (2000) The primary process and the unconscious: experimental evidence supporting two psychoanalytic presuppositions. Int. J. Psychoanal. 81: 553–569.

Brakel LAW (2004) The psychoanalytic assumption of the primary process: extra-psychoanalytic evidence and findings. J. American Psychoanal. Ass. 52: 1131–1161.

Broca P (1878) Anatomie comparée des circonvolutions cérébrales. Le grand lobe limbique et la scissure limbique dans la série des mammifères. Rev. Anthropol. 1, Ser. 2: 385–498.

Brown JW (1967) Physiology and phylogenesis of emotional expression. Brain Res. 5: 1–14.

Brown JW (1972) Aphasia, Apraxia, and Agnosia. Thomas, Springfield, Illinois.

Brown JW, Jaffe J (1975) Hypothesis on cerebral dominance. Neuropsychologia 13: 107–110.

Brown JW (1977) Mind, brain and consciousness: The neuropsychology of cognition. Academic Press, New York.

Brown JW (1978) Lateralisation: A brain model. Brain & Lang., 5: 258–261.

Brown JW (1979a) Language representation in the brain. In: Steklis H, Raleigh M (Eds) Neurobiology of social communication in primates. Academic Press, New York.

Brown JW (1979b) Thalamic mechanisms in language. In: Gazzaniga M (Ed.) Handbook of Neuropsychology, Plenum Press, New York pp 215–236.

Brown JW (1980) Brain structure and language production: a dynamic view. In: Caplan D (Ed.) Biological studies of mental processes. MIT Press, Cambridge, MA.

Brown JW (1982) Hierarchy and evolution in neurolinguistics. In: Arbib M, Caplan D, Marshall J (Eds) Neural models of language processes. Academic Press, New York.

Brown JW, Leader B, Blum C (1983) Hemiplegic writing in severe aphasia. Brain and Language 19: 204–215.

Brown JW (1983a) Microstructure of perception: physiology and patterns of breakdown. Cognition and Brain Theory 6: 145–184.

Brown JW (1983b) Emergence and time in microgenetic theory. J. American Acad. Psychoanal. 11: 35–54.

Brown JW (1985) Imagery and the microstructure ofperception. J. Neurolinguistics 1: 89–128.

Brown JW (1987) The microstructure of action. In: Perecman E. (Ed.) The frontal lobes re-visited. IRBN Press, New York.

Brown JW (1988a) The life of the mind: Selected papers. Lawrence Erlbaum Associates, Hillsdale, New Jersey.

Brown JW (1988b) Neuropsychology of visual perception. Lawrence Erlbaum Associates, Hillsdale, New Jersey.

Brown JW (1998) Psychoanalysis and process theory. Ann. N.Y. Acad. Sci. 843: 91–106.

Buchanan TW, Tranel D, Adolphs R (2005). Emotional autobiographical memories in amnesic patients with medial temporal lobe damage. J. Neurosci. 25: 3151–3160.

Burgess PW, Shallice T (1996) Bizarre responses, rule detection and frontal lobe lesions. Cortex 32: 241–259.

Buzsaki G (2002) Theta oscillations in the hippocampus. Neuron 33: 325–340.

Caine ED, Hunt RD, Weingartner H, Ebert MH (1978) Huntington's dementia. Clinical and neuropsychological features. Arch. Gen. Psychiatry 35: 377–384.

Calamandrei G, Keverne EB (1994) Differential expression of Fos protein in the brain of female mice dependent on pup sensory cues and maternal experience. Behav. Neurosci. 108: 113–120.

Camara Magalhaes S, Vitorino Souza C, Rocha Dias T, Felipe Carvalhedo de Bruin P, de Bruin VM (2005) Lifestyle regularity measured by the social rhythm metric in Parkinson's disease. Chronobiol Int. 22: 917–924.

Cannon WB (1953) Bodily changes in pain, hunger, fear and rage. 2nd ed. Borton, Boston.

Cascino GD, Sutula TP (1989) Thirst and compulsive water drinking in medial basal limbic epilepsy: an electroclinical and neuropathological correlation. J. Neurol. Neurosurg. Psychiatry 52: 680–681.

Ceccaldi M, Milandre L (1994) A transient fit of laughter as the inaugural symptom of capsular-thalamic infarction. Neurology 44: 1762.

Chee MWL, Kotagal P, Van Ness PC, Gragg L, Murphy D, Lüders HO. (1993) Lateralising signs in intractable partial epilepsy: blinded multiple-observer analysis. Neurology 43: 2519–2525.

Claparéde E (1911) Recognition et moiieté. Archives de Psychologie 11: 79–90.

Colbern D, Isaacson RL, Bohus B, Gispen WH (1977) Limbic-midbrain lesions and ACTH- induced excessive grooming. Life Sci. 21: 393–402.

Cory GA, Gardner R (2002) The evolutionary neuroethology of Paul MacLean. Convergences and Frontiers. Praeger Publishers, Westport, CT.

Creutzfeld OD (1983) Cortex Cerebri. Springer, Berlin.

Cummins DD (1996) Dominance hierarchies and the evolution of human reasoning. Minds and Machines 6: 463–480.

Damasio AR, Tranel D, Damasio H (1991) Somatic markers and the guidance of behaviour. In: Levin H, Eisenberg H, Benton A (Eds) Frontal lobe function and dysfunction. Oxford Univ. Press, New York, pp 217–228.

Darwin CR (1837) Notebook B (Transmutation of species).

Darwin CR (1859) On the origin of species by means of natural selection. John Murray, London.

Darwin CR (1871) The descent of man, and selection in relation to sex. John Murray, London.

Darwin CR (1872). The expression of the emotions in man and animals. John Murray, London.

Dehaene S, Changeux JP (2004) Neural mechanisms for access to consciousness. In: Gazzaniga MS (Ed) The Cognitive Neurosciences III. The MIT Press, Cambridge, London. pp 1145–1157.

de Waal FBM (1986) The integration of dominance and social bonding in primates. Q. Rev. Biol. 61: 459–479.

Dewhurst K (1982) Hughlings Jackson on psychiatry. Sanford, Oxford.

Dixon F (1971) Subliminal perception: The nature of a controversy. McGraw-Hill, London.

Dobesberger J, Walser G, Unterberger I, Embacher N, Luef G, Bauer G, Benke T, Bartha L, Ulmer H, Ortler M, Trinka E (2004) Genital automatisms: a video-EEG study in patients with medically refractory seizures. Epilepsia 45: 777–780.

Eibl-Eibesfeldt I (1971) !Ko-Buschleute (Kalahari) – Schamweisen und Spotten. Homo 22: 261–266.

Eibl-Eibesfeldt I (1984) Die Biologie des menschlichen Verhaltens. Grundriß der Humanethologie. Piper, München/Zürich.

Eibl-Eibesfeldt I (1999) Grundriß der vergleichenden Verhaltensforschung. Piper, München.

Enard W, Przeworski M, Fisher SE, Lai CSL, Wiebe V, Kitano T, Monaco AP, Paäbo S (2002) Molecular evolution of FOXP2, a gene involved in speech and language. Nature 418: 869–872.

Engel J, Jr. (1996) Introduction to the temporal lobe epilepsies. Epilepsy Res. 26: 141–150.

Engel J, Jr. Williamson PD, Wieser HG (1997) Mesial temporal lobe epilepsy. In: Engel J, Jr., Pedley TA (Eds) Epilepsy: A comprehensive textbook. Philadelphia: Lippincott-Raven Publishers.

Fish DR, Gloor P, Quesney FL, Olivier A (1993) Clinical responses to electrical brain stimulation of the temporal and frontal lobes in patients with epilepsy. Pathophysiological implications. Brain 116: 397–414.

Fishbein HD (1976) Evolution, development, and children's learning. Goodyear, Pacific Palisades, California.

Fisher C (1960) Subliminal and supraliminal influences on dreams. American J. Psychiatr. 116: 1009–1017.

Fleischer S, Slotnick BM (1978) Disruption of maternal behaviour in rats with lesions of the septal area. Physiol. Behav. 21: 189–200.

Franck D (1997) Verhaltensbiologie. 3. Auflage, Thieme Verlag, Stuttgart, New York, pp 11–14.

Freud S (1891) Zur Auffassung der Aphasien. Eine kritische Studie. Fischer Verlag, Frankfurt am Main 2001.

Freud S (1897) Brief 146. In: Masson JM (Ed) Sigmund Freud – Briefe an Wilhelm Fließ. S. Fischer. Frankfurt am Main, 1986, pp 301–305.

Freud S (1900a) Die Traumdeutung. Gesammelte Werke, II/III, Fischer Verlag, Frankfurt am Main 1999, pp 542–546.

Freud S (1900b) Die Traumdeutung. Gesammelte Werke, II/III, Fischer Verlag, Frankfurt am Main 1999, pp 620–622.

Freud S (1905a) Drei Abhandlungen zur Sexualtheorie. Gesammelte Werke, V, Fischer Verlag, Frankfurt am Main 1999, p 67.

Freud S (1905b) Drei Abhandlungen zur Sexualtheorie. Gesammelte Werke, V, Fischer Verlag, Frankfurt am Main 1999, pp 69–70.

Freud S (1905c) Drei Abhandlungen zur Sexualtheorie. Gesammelte Werke, V, Fischer Verlag, Frankfurt am Main 1999, p 98.

Freud S (1911a) Formulierungen über die zwei Prinzipien des psychischen Geschehens. Gesammelte Werke, VIII, Fischer Verlag, Frankfurt am Main 1999, pp 231–232.

Freud S (1911b) Psychoanalytische Bemerkungen über einen autobiographisch beschriebenen Fall von Paranoia, VIII, Fischer Verlag, Frankfurt am Main 1999, pp 319–320.

Freud S (1913a) Totem und Tabu. Gesammelte Werke, IX, Fischer Verlag, Frankfurt am Main 1999.

Freud S (1913b) Totem und Tabu. Gesammelte Werke, IX, Fischer Verlag, Frankfurt am Main 1999, pp 122–194.

Freud S (1915a) Das Unbewusste. Gesammelte Werke, X, Fischer Verlag, Frankfurt am Main 1999, pp 264–303.

Freud S (1915b) Übersicht der Übertragungsneurosen. Grubrich-Simitis I (Ed). S. Fischer Verlag, Frankfurt am Main 1985, pp 70–81.

Freud S (1923a) Das Ich und das Es. Gesammelte Werke, XIII, Fischer Verlag, Frankfurt am Main 1999, pp 235–290.

Freud S (1923b) Das Ich und das Es. Gesammelte Werke, XIII, Fischer Verlag, Frankfurt am Main 1999, p 244.

Freud S (1926) Hemmung, Symptom und Angst. Gesammelte Werke, XIV, Fischer Verlag, Frankfurt am Main 1999, pp 111–206.

Freud S (1930) Das Unbehagen in der Kultur. Gesammelte Werke, XIV, Fischer Verlag, Frankfurt am Main 1999, pp 421–506.

Freud S (1932) Neue Folge der Vorlesungen zur Einführung in die Psychoanalyse. Gesammelte Werke, XV, Fischer Verlag, Frankfurt am Main 1999, p 80.

Fried I, Wilson CL, MacDonald KA, Behnke EJ (1998) Electric current stimulates laughter. Nature 391: 650.

Fulton JF (1934) Forced grasping and groping in relation to the syndrome of the premotor area. Arch. Neurol. Psychiat. 31: 221–235.

Gajdusek DC (1970) Physiological and psychological characteristics of Stone Age man. In: Symposium on biological bases of human behaviour. Eng. Sci. 33, 26–33, 56–62.

Galin D (1974) Implications for Psychiatry of left and right cerebral specialisation: A neurophysiological context for unconscious processes. Arch. Gen. Psychiat. 31: 572–583.

Gallese V, Goldman A (1999) Mirror neurons and the simulation theory of mind-reading. Trends Cogn. Sci. 2: 493–500.

Gardner H, Ling PK, Flamm L, Silverman J (1975) Comprehension and appreciation of humorous material following brain damage. Brain 98: 399–412.

Gardner R (1982) Mechanisms in manic-depressive disorder. An evolutionary model. Arch. Gen. Psychiatry 39: 1436–1441.

Gardner R, Price JS (1999) Sociophysiology and depression. In: Joiner T, Coyne JC (Eds) The interactional nature of depression: Advances in interpersonal approaches, APA Books, Washington DC, pp 247–268.

Gedo JE, Goldberg A (1973) Models of the mind. A psychoanalytic theory. The University of Chicago Press, Chicago, London pp 73–150.

Goel V, Dolan RJ (2001) The functional anatomy of humour: segregating cognitive and affective components. Nature Neurosci. 4: 237–238.

Goldby F, Gamble HJ (1957) The reptilian cerebral hemispheres. Biol Rev. 32: 383–420.

Goldstein RG (1995) The higher and lower in mental life: an essay on J. Hughlings Jackson and Freud. J. Amer. Psychoanal. Assn. 43: 495–515.

Gould SJ (1977) Ontogeny and Phylogeny. The Belknap Press of Harvard University Press, Cambridge, MA; London, England.

Graybiel AM (2008) Habits, rituals, and the evaluative brain. Annu. Rev. Neurosci. 31: 359–387.

Greenberg N, Ferguson JL, MacLean PD (1976) A neuroethological study of display behaviour in lizards. Neuroscience 2: 689.

Greenberg N (1978) Ethological considerations in the experimental study of lizard behaviour. In: Greenberg N, MacLean PD, (Eds.) The behaviour and neurology of lizards. U.S. Government Printing Office, Washington, D.C., pp 203–224.

Greenberg N, MacLean PD, Ferguson JL (1979) Role of the paleostriatum in species-typical behaviour of the lizard *Anolis Carolinensis*. Brain Res. 172: 229–241.

Groddeck G (1923) Das Buch vom Es. Fischer Verlag, Frankfurt am Main 1979, p 7.

Haeckel E (1866) Die Generelle Morphologie der Organismen. Berlin, G. Reimer.

Haeckel E (1874) Anthropogenie oder Entwicklungsgeschichte des Menschen. Engelmann, Leipzig.

Halgren E, Walter RD, Cherlow DG, Crandall PH (1978) Mental phenomena evoked by electrical stimulation of the human hippocampal formation and amygdala. Brain 101: 83–117.

Hanlon R (1990) Cognitive microgenesis: A neuropsychological perspective. Springer-Verlag, New York.

Harris VA (1964) The life of the Rainbow Lizard. Hutchinson, London.

Hassler R (1961) Motorische und sensible Effekte umschriebener Reizungen und Ausschaltungen im menschlichen Zwischenhirn. Dtsch. Z. Nervenheilkd. 183: 148– 171.

Heath RG (1963) Electrical self-stimulation of the brain in man. Am. J. Psychiatry 120: 571–577.

Hebben N, Corkin S, Eichenbaum H, Shedlack K (1985) Diminished ability to interpret and report internal states after bilateral medial temporal resection: case H.M. Behav. Neurosci. 99: 1031–1039.

Hernandez-Peon R (1966) A neurophysiological model of dreams and hallucinations. J. Nerv. Ment. Dis. 141: 623–650.

Hess WR (1949) Das Zwischenhirn. Syndrome, Lokalisationen, Funktionen. Schwabe, Basel.

Hornak JE, Rolls T, Wade T (1996) Face and voice expression identification in patients with emotional and behavioural changes following ventral frontal lobe damage. Neuropsychologia 34: 247–261.

Hutchison WD, Davis KD, Lozano AM et al. (1999) Pain-related neurons in the human cingulate cortex. Nat. Neurosci. 2: 403–405.

Jackson JH (1931) Taylor J. (Ed) Selected writings of John Hughlings Jackson. Staples Press, London.

Jackson SW (1969) The history of Freud's concepts of regression. J. American Psycho- anal. Assoc. 17: 743–784.

Jakobsson J, Cordero MI, Bisaz R, Groner AC, et al. (2008) KAP1-mediated epigenetic repression in the forebrain modulates behavioural vulnerability to stress. Neuron 60: 818–831.

Jansky J, Ebner A, Szupera Z, Schulz R, Hollo A, Szücs A, Clemens B (2004) Orgasmic aura: a report of seven cases. Seizure 13: 441–444.

Jerison HJ (1973) Evolution of the brain and intelligence. Academic Press, New York and London.

Jog MS, Kubato Y, Connolly CI, Hillegaart V, Graybiel AM (1999) Building neural representations of habits. Science 286: 1745–1749.

Kallen B (1951) On the ontogeny of the reptilian forebrain. Nuclear structures and ventricular sulci. J Comp. Neurol. 95: 307–347.

Kandel ER (2006) Psychiatrie, Psychoanalyse und die neue Biologie des Geistes. Suhrkamp, Frankfurt am Main, pp 119–183.

Kaplan-Solms K, Solms M (2000) Clinical studies in Neuro-Psychoanalysis. Karnac Books, London.

Karnath HO, Wallesch CW, Zimmermann P (1991) Mental planning and anticipatory processes with acute and chronic frontal lobe lesions: a comparison of maze performance in routine and non-routine situations. Neuropsychologia 29: 271–290.

Karnath HO, Thier P (2003) Neuropsychologie. Springer Verlag, Berlin, Heidelberg, New York.

Kataoka S, Hori A, Shirakawa T, Hirose G (1997) Paramedian pontine infarction. Neurological/topographical correlation. Stroke 28: 809–815.

Kennard C (1989a) Hierarchies and the visual system. In: Kennard C, Swash M (Eds) Hierarchies in neurology. A reappraisal of a Jacksonian concept. Springer-Verlag, London, New York, pp 87–92.

Kennard C (1989b) Hierarchical aspects of eye movement disorders. In: Kennard C, Swash M (Eds) Hierarchies in neurology. A reappraisal of a Jacksonian concept. Springer–Verlag, London, New York, pp 151–158.

Kim JS, Choi-Kwon KS (2000) Poststroke depression and emotional incontinence: correlation with lesion location. Neurology 54: 1805–1810.

Kornhuber HH, Deecke L (1965) Hirnpotentialänderungen bei Willkürbewegungen und passiven Bewegungen des Menschen: Bereitschaftspotential und reafferente Potentiale. Pflügers Arch. 281: 1–17.

Krolak-Salmon P, Henaff MA, Vighetto A, Bauchet F, Bertrand O, Maugiere F et al (2006) Experiencing and detecting happiness in humans: The role of the supplementary motor area. Ann. Neurol. 59: 196–199.

Lamarck JB (1801) Systéme des animaux sans vertébres précédé du „Discours d'ouverture du cours de zoologie de l'an VIII“ (Paris, an IX).

Lamarck JB (1802) Recherches sur l'organisation des corps vivans précédé du „Discours d'ouverture du cours de zoologie de l'an X“ (Paris, an X).

Lamarck JB (1809) Philosophie zoologique. 2 vols, Paris.

Lamarck JB (1815–1822) Histoire naturelle des animaux sans vertèbres. 7 vols, Paris.

Lang EM, Schenkel R (1960–61) Goma, das Basler Gorillakind. Documenta Geigy, Basel.

Leutmezer F, Serles W, Bacher J, Gröppel G, Pataraia E, Aull S, Olbrich A, Czech T, Baumgartner C (1999) Genital automatisms in complex partial seizures. Neurology 52: 1188–1191.

Lewes GH (1879) The study of psychology. Houghton, Boston.

Loch W (1999) Die Krankheitslehre der Psychoanalyse. S. Hirzel Verlag, Stuttgart, Leipzig, pp 25–30.

Loddenkemper T, Foldvary N, Raja S, Neme S, Lüders HO (2003) Ictal urinary urge: further evidence for lateralisation to the non-dominant hemisphere. Epilepsia 44: 124–126.

Lorenz K (1937) Bildung des Instinktbegriffes. Naturwissenschaften 25: 289–300.

Lorenz K (1978) Vergleichende Verhaltensforschung. Grundlagen der Ethologie. Springer, Wien.

Luciano D, Devinsky O, Perrine K (1993) Crying seizures. Neurology 43: 2113–2117.

MacLean PD (1952) Some psychiatric implications of physiological studies on frontotemporal portion of limbic system (visceral brain). Electroencephalogr. Clin. Neurophysiol. 4: 407–418.

MacLean PD (1954) The limbic system and its hippocampal formation. Studies in animals and their possible application to man. J Neurosurg. 11: 29–44.

MacLean PD (1957) Visceral functions of the nervous system. Annu. Rev. Physiol. 19: 397–416.

MacLean PD (1962) New findings relevant to the evolution of psychosexual functions of the brain. J. Nerv. Ment. Dis. 135: 289–301.

MacLean PD (1964) Mirror display in the squirrel monkey *Saimiri sciureus.* Science 146: 950–952.

MacLean PD (1969) The internal-external bonds of the memory process. J. Nerv. Ment. Dis. 149: 40–47.

MacLean PD (1970) The triune brain, emotion, and scientific bias. In: Schmitt FO (Ed) The Neurosciences. Second study program. Rockefeller University Press, New York, 336–349.

MacLean PD (1973) The brain's generation gap: Some human implications. Zygon J. Relig. Sci. 8, 113–127.

MacLean PD (1975a) On the evolution of three mentalities. Man-Environment-Systems 5: 213–224.

MacLean PD (1975b) The imitative-creative interplay of our three mentalities. In: Harris H (Ed.) Astride the two cultures: Arthur Koestler at 70, Hutchinson, London, pp 187–211.

MacLean PD (1975c) An ongoing analysis of hippocampal inputs and outputs: Microelectrode and anatomic findings in squirrel monkeys. In: Isaacson RI, Pribram KH (Eds) The Hippocampus, Volume 1, Plenum Press, New York, pp 177–211.

MacLean PD (1977) An evolutionary approach to brain research on prosematic (nonverbal) behaviour. In: Rosenblatt JS and Komisaruk BR (Eds) Reproductive Behaviour and Evolution. Plenum Press, New York, pp 177–211.

MacLean PD (1978) Effects of lesions of globus pallidus on species-typical display behaviour of squirrel monkeys. Brain Res. 149: 175–196.

MacLean PD (1985) Brain evolution relating to family, play, and the separation call. Arch. Gen. Psychiatry 42: 405–417.

MacLean PD (1987) The midline frontolimbic cortex and the evolution of crying and laughter. In: Perecman E (Ed) The frontal lobes revisited. IRBN Press, New York, pp 121–140.

MacLean PD, Newman JD (1988) Role of midline frontolimbic cortex in production of the isolation call of squirrel monkeys. Brain Res. 450: 111–123.

MacLean PD (1990) The Triune Brain in evolution. Role in paleocerebral functions. Plenum Press, New-York, London.

Mahner M, Bunge M (2000) Philosophische Grundlagen der Biologie. Springer Verlag, Berlin.

Malthus TR (1798) An essay on the principle of population. Johnson J, St. Paul's Church Yard, London.

Marín O, Smeets WJAJ, Gonzáles A (1998) Basal ganglia organisation in amphibians: evidence for a common pattern in tetrapods. Prog. Neurobiol. 55: 363–397.

Mayr E (1997) This is biology. Harvard University Press, Cambridge.

McGaugh JL (2004) The amygdala modulates the consolidation of memories of emotionally arousing experiences. Annu. Rev. Neurosci. 27: 1–28.

McLaughlin JT (1978) Primary and secondary process in the context of cerebral hemispheric specialisation. Psychoanalytic Q. 47: 237–266.

Medin D (1990) Similarity involving attributes and relations: judgements of similarity and difference are not inverse. Psychol. Sci. 1: 64–69.

Meerlo JAM (1962) The dual meaning of human regression. Psychoanal. Rev. 49: 77–86.

Mellars P (1998) Neanderthals, modern humans and the archaeological evidence for language. In: Jablonski NG, Aiello LC (Eds) The origin and diversification of language. Academy of Sciences, San Francisco.

Mendez MF, Nakatawase TV, Brown CV (1999) Involuntary laughter and inappropriate hilarity. J. Neuropsychiatry Clin. Neurosci. 11: 253–258.

Montagu A (1959) Natural selection and the origin and evolution of weeping in man. Science 130: 1572–1573.

Morgane PJ, Galler JR, Mokler DJ (2005) A review of systems and networks of the limbic forebrain/limbic midbrain. Progr. Neurobiol. 75: 143–160.

Mullan S, Penfield W (1959) Illusions of comparative interpretation and emotion; production by epileptic discharge and by electrical stimulation in the temporal cortex. AMA Arch. Neurol. Psychiatry 81: 269–284.

Murphy MR, MacLean PD, Hamilton SC (1981) Species-typical behaviour of hamsters deprived from birth of neocortex. Science 213: 459–461.

Newman JD (1985) The infant cry of primates. An evolutionary perspective. In: Lester BM, Boukydis CFZ (Eds), Infant Crying: Theoretical and Research Perspectives. Plenum Press, New York.

Nieuwenhuys R, Voogd J, van Huijzen C (2007) The human central nervous system. 4th Ed. Springer.

Noble GK, Bradley HT (1933) The mating behaviour of lizards: its bearing on the theory of sexual selection. Ann N.Y Acad. Sci. 35: 25–100.

Owren MJ, Bachorowski JA (2003) Reconsidering the evolution of non-linguistic communication: the case of laughter. J. Nonverbal Behav. 27: 183–197.

Panksepp J (1998) Affective Neuroscience: The foundations of human and animal emotions. Oxford University Press, New York and Oxford.

Panksepp J (2005) Beyond a joke: from animal laughter to human joy. Science 308: 62–63.

Parker GA (1974) Assessment strategy and the evolution of fighting behaviour. J. Theoret. Biol. 47: 223–243.

Parvizi J, Anderson SW, Martin CO, Damasio H, Damasio AR (2001) Pathological laughter and crying. A link to the cerebellum. Brain 124: 1708–1719.

Peiper A, Thomas H (1952) Leerlaufendes Brustsuchen. Mschr Kinderheilk. 101: 303–307.

Penfield W, Welch K (1951) The supplementary motor area of the cerebral cortex. Arch. Neurol. Psychiatry 66: 289–317.

Perecman E (1985) Ukeleles, Condessors, and Fosetch. Language Sciences 7: 177–214.

Pilleri G (1960a) Kopfpendeln („leerlaufendes Brustsuchen“) bei einem Fall von Pick'scher Krankheit. Arch. Psychiat. Nervenkr. 200: 603–611.

Pilleri G (1960b) Über das Auftreten von Kletterbewegungen im Endstadium eines Falles von Morbus Alzheimer. Arch. Psychiat. Nervenkr. 200: 455–461.

Pilleri G (1961) Über das chronologische Auftreten von motorischen Schablonen des Oralsinnes, deren ontogenetische Bedeutung und klinisch-anatomische Zusammenhänge bei atrophisierenden Hirnerkrankungen. Schweiz. Arch. Neurol. Neurochir. Psych. 88: 273–298.

Pilleri G (1971) Instinktbewegungen des Menschen in biologischer und neuropathologischer Sicht. Akt. Fragen Psychiat. Neurol. 11: 1–37.

Pilleri G, Poeck K (1964) Arterhaltende und soziale Instinktbewegungen als neurologische Symptome beim Menschen. Psychiat. Neurol. 147: 193–238.

Ploog DW, MacLean PD (1963) Display of penile erection in squirrel monkey (*Saimiri sciureus*). Anim. Behav. 11: 32–39.

Ploog DW (2003) The place of the triune brain in psychiatry. Physiology & Behavior 79: 487–493.

Plotnik R (1968) Changes in social behaviour of squirrel monkeys after anterior temporal lobectomy. J. Comp. Physiol. Psychol. 66: 369–377.

Poeck K, Hubach H (1963) Rhythmische orale Automatismen bei Dezerebrationszuständen. Deutsch. Z. Nervenheilk. 185: 37–52.

Poeck K, Pilleri G (1963) Pathologisches Lachen und Weinen. Schweiz. Arch. Neurol. Neurochir. 92: 323–370.

Poeck K (1969) Pathophysiology of emotional disorders associated with brain damage. In: Vinkens PJ, Bruyn GW (Eds) Handbook of clinical neurology. Elsevier, Amsterdam pp 343–367.

Poeck K (1985) Pathological laughter and crying. In: Vinken PJ, Bruyn GW, Klawans HL (Eds) Handbook of clinical neurology, vol 1: Clinical neuropsychology. Elsevier, Amsterdam pp 257–263.

Pötzl O (1917) Experimentell erregte Traumbilder und ihre Beziehungen zum indirekten Sehen. Zeitschr. Ges. Neurol. Psychiatr. 37: 278–349.

Pontius AA (1993) Neuroethological aspects of certain limbic seizure-like dysfunctions: Exemplified by limbic psychotic trigger reaction (motiveless homicide with intact memory). Integrative Psychiatry 9: 151–167.

Pontius AA (2002) Neuroethology, exemplified by limbic seizures with motiveless homicide in "Limbic Psychotic Trigger Reaction". In: Cory GA, Gardner R (Eds) The evolutionary neuroethology of Paul MacLean. Convergences and Frontiers. Praeger Publishers, Westport, CT, pp 167–191.

Prechtl HFR, Schleidt WM (1950) Auslösende und steuernde Mechanismen des Saugaktes. Z. vgl. Physiol. 32: 257–262.

Prechtl HFR (1953) Über die Koppelung von Saugen und Greifen beim Säugling. Naturwissenschaften 40: 347.

Preuss TM (2004) What is it like to be a human? In: Gazzaniga MS (Ed) The cognitive neurosciences III. The MIT Press, Cambridge, London, pp 5–22.

Price J (1967) The dominance hierarchy and the evolution of mental illness. Lancet 2: 243–246.

Price J, Sloman L, Gardner R, Gilbert P, Rohde P (1994) The social competition hypothesis of depression. British J. Psychiatry 164: 309–315.

Price J (2002) The triune brain, escalation, de-escalation strategies, and mood disorders. In: Cory GA, Gardner R (Eds) The evolutionary neuroethology of Paul MacLean. Convergences and Frontiers. Praeger Publishers, Westport, CT, pp 107–117.

Rado S (1969) Adaptational psychodynamics: motivation and control. Science House, New York.

Ramachandran VS (1998) The neurology and evolution of humour, laughter and smiling: The false alarm theory. Med. Hypotheses 51: 351–354.

Ramachandran VS (2003) Humour and laughter: A biological hypothesis. In: Baars BJ, Banks WP, Newman JB (Eds) Essential sources in the scientific study of consciousness. The MIT Press, Cambridge, London, pp 819–820.

Ramón y Cajal S (1892) El nuevo concepto de la histologia de los centros nervi-osos. III. Corteza gris del cerebro. Rev. Cien. Med. Barcelona 18: 457–476.

Ramón y Cajal S (1894) Les nouvelles idées sur la structure du système nerveux chez l'homme et chez les vertébres. Reinwald, Paris.

Ramón y Cajal S (1909–1911) Histologie du système nerveux de l'homme et des vertébrés. 2 Vol. Maloine, Paris.

Rapaport D (1970) Die Struktur der psychoanalytischen Theorie. Ernst Klett Verlag, Stuttgart, pp 43–77.

Ritvo LB (1965) Darwin as the source of Freud's Neo-Lamarckianism. J. Amer. Psychoanal. Assn. 18: 195–208.

Ritvo LB (1993) Darwin's influence on Freud: A tale of two sciences. Yale Univ. Press, New Haven & London

Rolls ET, Thorpe SJ, Maddison S, Roper-Hall A, Puerto A, Perret D (1979) Activity of neurons in the neostriatum and related structures in the alert animal. In: Divac I and Oberg RGE (Eds) The Neostriatum. Pergamon Press, Elmsford, NY, pp 163–182.

Roth G, Wullimann MF (1996) Evolution der Nervensysteme und der Sinnesorgane. In: Dudel J, Menzel R, Schmidt RF (Eds) Neurowissenschaft. Vom Molekül zur Kognition. Springer Verlag, Berlin, Heidelberg, New York, pp 1–31.

Saka E, Goodrich C, Harlan P, Madras BK, Graybiel AM (2004) Repetitive behaviours in monkeys are linked to specific striatal activation patterns. J. Neurosci. 24: 7557–7565.

Salk L (1960) The effects of the normal heartbeat sound on the behaviour of the newborn infant: Implications for mental health. World Ment. Health 12: 168–175.

Sander F (1928) Experimentelle Ergebnisse der Gestaltpsychologie. In: Becher E (Ed) 10. Kongressbericht experimentelle Psychologie. Fischer, Jena.

Sander F (1930) Structures, totality of experience, and Gestalt. In: Murchinson C (Ed) Psychologies of 1930. Clark University Press, Worcester, MA, pp 188–204.

Schepelmann F (1979) Rhythmic patterns of motor activity after lesions of the central nervous system in man. Acta Neurochirurgica 49: 153–189.

Schjelderup-Ebbe (1922) Beiträge zur Sozialpsychologie des Haushuhns. Zeitschrift für Psychologie 88: 225–252.

Schott JM, Rossor MN (2002) The grasp and other primitive reflexes. J Neurol Neurosurg Psychiatry 74: 558–560.

Schreiner L, Kling A (1956). Rhinencephalon and behaviour. Amer. J. Physiol. 184: 486–490.

Schur M (1960) Phylogenesis and ontogenesis of affect- and structure-formation and the phenomenon of repetition compulsion. Int. J. Psychoanal. 41: 275–287.

Seiss R (1965) Beobachtungen zur Frage der Übersprungsbewegungen im menschlichen Verhalten. Psychol. Beitr. 8: 1–97.

Shamay-Tsoory SG, Tomer R, Berger BD, Aharon-Peretz J (2003) Characterisation of empathy deficits following prefrontal brain damage: The role of the right ventromedial prefrontal cortex. J. Cogn. Neurosci. 15: 324–337.

Shammi P, Stuss DT (1999) Humour appreciation: a role of the right frontal lobe. Brain 122: 657–666.

Shevrin H, Luborsky L (1960) The rebus technique: a method for studying primary process transformations of briefly exposed pictures. J. Nerv. & Ment. Dis. 133: 479–488.

Shevrin H, Fisher C (1967) Changes in the effects of a waking subliminal stimulus as a function of dreaming and non-dreaming sleep. J. Abnormal Psychol. 72: 362–368.

Shevrin H, Dickman S (2003) The psychological unconscious: a necessary assumption for all psychological theory? In: Baars BJ (Ed) Essential sources in the scientific study of consciousness. The MIT Press, Cambridge, London.

Singer HS, Reiss AL, Brown JE et al. (1993) Volumetric MRI changes in basal ganglia of children with Tourette syndrome. Neurology 43: 950–956.

Slotnick BM, Nigrosh BJ (1975) Maternal behaviour in mice with cingulate cortical, amygdala, or septal lesions. J. Comp. Physiol. Psychol. 88: 118–127.

Smith E, Medin D (1981) Categories and Concepts. Harvard Univ. Press, Cambridge, MA.

Solms M, Saling M (1986) On psychoanalysis and neuroscience: Freud's attitude to the localisationist tradition. Int. J. Psychoanal. 67: 397–416.

Solms M, Turnbull O (2002a) The brain and the inner world. An introduction to the neuroscience of subjective experience. Karnac, London – New York, pp 115–119.

Solms M, Turnbull O (2002b) The brain and the inner world. An introduction to the neuroscience of subjective experience. Karnac, London – New York, pp 101–104.

Spencer H (1855) The principles of psychology. London: Longman, Brown, Green, and Longmans.

Spencer H (1896) The principles of psychology, 3rd edition, 2 volumes in 3. Appleton, New York.

Spitz RA (1965) Vom Säugling zum Kleinkind (11.Auflage, 1987). Klett-Cotta, Stuttgart.

Stevens A, Price J (2000) Evolutionary Psychiatry: A new beginning (Second Edition). Routledge, London, New York.

Striedter GF (2005) Principles of brain evolution. Sinauer Associates, Sunderland, MA.

Stroud J (1967) The fine structure of psychological time. Annals N. Y. Acad. Sci. 138: 623–631.

Stroufe LA, Waters E (1976) The ontogenesis of smiling and laughter: A perspective on the organisation of development in infancy. Psych. Rev. 83: 173–189.

Sulloway F (1979) Freud: Biologist of the mind: Beyond the psychoanalytic legend. Basic Books, New York, pp 361–415.

Swanson LW (1987) Limbic system. In: Adelman G. (Ed.). Encyclopedia of Neuroscience. Birkhäuser, Boston, Basel, Stuttgart, 589–591.

Swash M (1989) Order and disorder in the motor system. In: Kennard C, Swash M (Eds) Hierarchies in neurology. A reappraisal of a Jacksonian concept. Springer-Verlag, London, New York, pp 113–122.

Tarpy R (1977) The nervous system and emotion. In: Candland DK, Fell JP, Keen E, Leshner AI, Tarpy R, Plutchik R (Eds) Emotion. Brooks/Cole, Monterey, California.

Taylor AE, Saint-Cyr JA (1995) The neuropsychology of Parkinson's disease. Brain Cogn. 28: 281–296.

Taylor J. (ed) (1931/32) Selected writings of John Hughlings Jackson, vols 1 and 2. Hodder and Stoughton, London. Reprinted (1958) Basic Books, New York.

Tinbergen N (1940) Die Übersprungsbewegung. Z. Tierpsychol. 4: 1–40.

Tinbergen N (1951) The study of instinct. Oxford University Press, London.

Trinka E, Walser G, Unterberger I, Luef G, Benke T, Bartha L, Ortler M, Bauer G (2003) Peri-ictal water drinking lateralizes seizure onset to the nondominant temporal lobe. Neurology 60: 873–876.

Urbach E, Wiethe C (1929) Lipoidosis cutis et mucosae. Virchows Arch. 273: 285–319.

Vallbo AB (1989) Single fibre microneurography and sensation. In: Kennard C, Swash M (Eds) Hierarchies in neurology. A reappraisal of a Jacksonian concept. Springer-Verlag, London, New York, pp 93–109.

Van Buren JM, Li CL, Ojemann GA (1966) The fronto-striatal arrest response in man. Electroencephalogr. Clin Neurophysiol. 21: 114–130.

Vogt C, Vogt O (1920) Lehre der Erkrankungen des striären Systems. J. Psychol. Neurol 25: 628–846.

Wechsler JS, Bieber J, Balser GH (1936) Postural reflexes in patients with lesions of the frontal lobe. Arch. Neurol. Psychiat. 35: 1208–1215.

Weiller C, Chollet F, Friston KJ, et al. (1992) Functional reorganization of the brain in recovery from striatocapsular infarction in man. Ann Neurol 31: 463–472

Weinstein EA, Kahn RL (1955) Denial of illness: symbolic and physiological aspects. Charles C. Thomas Publisher, Springfield, Illinois.

Weiss P (1941) Self-differentiation of the basic patterns of coordination. Comp. Psychol. Monogr. 17: 1–96.

Werner H (1940a) Musical „micro-scales“ and „micromelodies“. Journal of Psychology 10: 149–156.

Werner H (1940b) Comparative psychology of mental development. Harper, New York.

Werner H (1956) Microgenesis and aphasia. Journal of Abnormal & Social Psychology 52: 347–353.

von Wickler W (1966) Ursprung und biologische Deutung des Genitalpräsentierens männlicher Primaten. Z. Tierpsychol. 23: 422–437.

Wieser HG (1983) Electroclinical features of the psychomotor seizure. Gustav Fischer, Stuttgart- New York.

Wieser S (1957a) Pathologie und Physiologie des Greifens. Fortschr. Neurol. Psychiat. 25: 317–341.

Wieser S (1957b) Schlüsselreize raumorientierender Zuwendereaktionen. Arch. Psychiat. Nervenkr. 195: 373–382.

Wiest G, Lehner-Baumgartner E, Baumgartner C (2006) Panic attacks in an individual with bilateral selective lesions of the amygdala. Arch. Neurol. 63: 1798–1801.

Wild B, Rodden FA, Grodd W, Ruch W (2003) Neural correlates of laughter and humour. Brain 126: 2121–2138.

Williams D (1956) The structure of emotions reflected in epileptic experiences. Brain 79: 29–67.

Witt K, Nuhsman A, Deuschl G (2002) Dissociation of habit-learning in Parkinson's and cerebellar disease. J. Cogn. Neurosci. 14: 493–499.

Wojtecki L, Nickel J, Timmermann L, Maarouf M, Südmeyer M et al. (2007) Pathological crying induced by deep brain stimulation. Mov. Disord. 22: 1314–1316.

Wortis H, Maurer WA (1942) Sham rage in man. Amer. J. Psychiat. 98: 638.

Yakovlev PI (1948) Motility, behaviour, and the brain. Stereodynamic organisation and neural coordinates of behaviour. J. Nerv. Ment. Dis. 107: 313–335.

Yin HH, Knowlton BJ (2006) The role of the basal ganglia in habit formation. Nature Rev. Neurosci 7: 464–476.
York GK, Steinberg DA (1995) Hughlings Jackson's theory of recovery. Neurology 45: 834–838.
Zeilig G, Drubach M, Katz-Zeilig M, Karatinos J (1996) Pathological laughter and crying in patients with closed traumatic brain injury. Brain Inj. 10: 591–597.

Index

G. Wiest, *Hierarchies in the Brain, Mind and Behaviour*,
https://doi.org/10.1007/978-3-031-78030-1

D

Q

R

S

The manufacturer's authorised representative in the EU is Springer Nature Customer Service Centre GmbH, Europaplatz 3, 69115 Heidelberg, Germany. If you have any concerns regarding our products, please contact ProductSafety@springernature.com

Printed and bound by CPI Group (UK) Ltd, Croydon, CR0 4YY
30/09/2026
02225591-0001